Nuclear Physics 3

I dedicate this book to my
darling baby Mariama

Nuclear Physics 3

Radiopharmaceuticals used in Nuclear Medicine

Ibrahima Sakho

ISTE

WILEY

First published 2024 in Great Britain and the United States by ISTE Ltd and John Wiley & Sons, Inc.

ISTE Ltd
27-37 St George's Road
London SW19 4EU
UK

www.iste.co.uk

John Wiley & Sons, Inc.
111 River Street
Hoboken, NJ 07030
USA

www.wiley.com

Library of Congress Control Number: 2024934862

British Library Cataloguing-in-Publication Data
A CIP record for this book is available from the British Library
ISBN 978-1-78630-993-8

Contents

Preface

Radionuclides are useful in many areas of everyday life: archaeology, biology, agronomy, medicine, industry etc. The most spectacular applications include radiochronometry (dating of archaeological objects, sediments and soils for the detection of anthropogenic pollutants, etc.) and nuclear medicine (radiopharmaceuticals used in nuclear medicine imaging, radiotherapy, etc.). The production of electrical energy in nuclear power plants exploits the properties of nuclear fission reactions. *Nuclear Physics 2* comprises four chapters devoted respectively to a description of the Big Bang model, a study of the different nucleosynthesis processes, the study of radiochronometers applied to dating and general information on radiopharmaceuticals used in nuclear medicine imaging.

This book, entitled *Nuclear Physics 3: Radiopharmaceuticals used in Nuclear Medicine*, consists of three chapters.

Chapter 1 is devoted to the study of *radiological imaging methods*. It begins with a presentation of medical imaging methods and their various medical imaging modalities. It then presents radiological imaging methods such as X-rays, γ-scan (PET) MRI scintigraphy and ultrasound. Next, photoelectric absorption, Compton effect and pair creation coefficients are introduced to express the total attenuation coefficient. The Hounsfield number is expressed as a function of the linear attenuation coefficient of water. The study then focuses on the treatment planning system (TPS), which is used to determine the dose distribution in tumors and surrounding areas during radiotherapy. Following this introduction, the principles of X-ray production, X-ray imaging and computed tomography (CT) or scanner are explained. The computer processing of elementary images to reconstruct the final image of each organ through which X-rays are passed features prominently in this section. The study then turns to nuclear magnetic resonance (NMR), explaining the principle of NMR and its physical model in biological tissues, the quantum model of Larmor precession, the phenomena of excitation and magnetic resonance, and the quantum model of the magnetic resonance phenomenon. The study then turns to magnetic resonance imaging (MRI), with an introduction of the two MRI techniques, closed-field MRI and open-field MRI, T_1 and T_2 relaxation times, spin–lattice and spin–spin relaxation phenomena, and the phenomenon of the free induction decay (FID) signal according to an exponential in T_2^* and T_2. Following these studies, the NMR imaging sequence – which is a chronological sequence of θ-angle radiofrequency pulses and magnetic field gradients – is introduced. This enables us to study in detail the two main methods of medical imaging sequences: the spin echo sequence and the gradient echo sequence. T_1, T_2 and proton density weighting of MRI images, as well as the influence of TR repetition time and TE echo time on an NMR imaging sequence, are explained in detail in this section. This is followed by a description of slice planes, slice selection gradient, frequency encoding gradient, phase encoding gradient, Fourier plane, MRI signal decoding and the ***principle*** of three-dimensional image acquisition using a triple Fourier transform (3DFT). The chapter concludes with an explanation of the principle of ultrasound imaging and the principle of Doppler ultrasound applied to the exploration of blood flow in a vessel.

Chapter 2 is devoted to the study of *technetized radiopharmaceuticals and radiothallium-201 used in nuclear medicine*. The chapter begins with a presentation of the main characteristics of the tracers MIBI ^{99m}Tc (MIBI = 2-methoxy-isobutyl-isonitrile), Tetrofosmin ^{99m}Tc and ^{201}Tl used in myocardial scintigraphy. The characteristics and dosimetry of conventional gamma cameras and latest-generation CZT (Cadmium Zinc Tellurium) gamma cameras are then presented. The notions of absorbed dose (expressed in gray, symbol Gy), equivalent dose (expressed in sieverts, symbol Sv) and weighting factor (without unit) are then defined. The advantages of tracers technetized on thallium-201 are examined, as are the uses of metastable technetium-99 in scintigraphy. For each of the radiopharmaceuticals or radiotracers considered, dosimetry is presented in tabular form for the various specific organs diagnosed, and the mechanisms by which ^{99m}Tc and ^{201}Tl are excreted from the body are described. In addition, the production chains for metastable technetium-99 and radiothallium-201 are explained, as well as the relationship between effective, biological and physical half-lives, and the physico-chemical and pharmacological properties of each radiotracer. This chapter describes the principle of ^{99m}Tc MIBI and ^{99m}Tc-labelled tetrofosmin scintigraphy, and radiopharmaceuticals used to diagnose or localize myocardial infarction or myocardial ischemia. Myocardial ^{99m}Tc MIBI scintigraphy, coupled with pharmacological stimulation sensitized by Dipyridamole (Persantine®), is also described. Following on from this development, the principle of lung ventilation scintigraphy using Technetium-99m-labeled diethylenetriamine pentaacetic acid (DTPA), denoted as [^{99m}Tc]-DTPA, and the principle of lung perfusion scintigraphy using a radiotracer consisting of ^{99m}Tc-labeled human macro-albumin aggregates (MAA), denoted as ^{99m}Tc-MAA, are studied. The study then focuses on the principle of brain perfusion scintigraphy with the radiopharmaceuticals ^{99m}Tc-HMPAO

(^{99m}Tc-hexamethylpropylene-amine-oxime) and ^{99m}Tc-ECD (^{99m}Tc-ethylenediyl bis-L-cysteine-diethyl-ester), which are used for the accurate diagnosis of cognitive disorders, strokes and epilepsy. The principle of ^{201}Tl myocardial scintigraphy, used without a vector to assess coronary perfusion, is then studied. Next, the procedure for myocardial perfusion scintigraphy is described in detail, consisting of an exercise test on a bicycle (cycle ergometer) or a treadmill, or a stress test. Following on from this, the principle of bone scintigraphy using ^{99m}Tc-labeled biphosphonate carrier molecules (BPs), including ^{99m}Tc-methyl diphosphonate (^{99m}Tc-MDP) is explored. Finally, we describe the principle of static renal scintigraphy using the radiopharmaceutical ^{99m}Tc-DMSA (^{99m}Tc-dimercapto succinic acid) to assess kidney morphology, the principle of dynamic renal scintigraphy using ^{99m}Tc-MAG$_3$ (^{99m}Tc-*mercapto acetyl triglycine*) to assess kidney function and the principle of gastric scintigraphy based on the use of a standardized low-fat meal consisting of two eggs labeled with ^{99m}Tc-radiolabeled colloidal rhenium sulfide.

Chapter 3 is dedicated to the study of *the radioisotopes fluorine-18, metastable krypton-81 and iodine-123, 125 and 131 used in nuclear medicine*. It begins with a study of the physico-chemical and pharmacological properties of the radiopharmaceutical 2-deoxy-2-[^{18}F]fluoro-D-glucose or ^{18}FDG, the most widely used PET radiopharmaceutical. The process of obtaining radiofluorine-18 via the reactions ^{20}Ne $(d, \alpha)^{18}$F and ^{18}O $(p, n)^{18}$F, the radiofluorine-18 decay scheme, the ^{18}FDG synthesis process, the principle of ^{18}FDG scintigraphy and the mechanism of ^{18}F excretion from the body are presented. Next, the physico-chemical and pharmacological properties of metastable krypton-81 (^{81m}Kr) are described, along with the ^{81m}Kr tracer production chain using the ^{81}Rb/^{81m}Kr generator and the principles of ^{81m}Kr ventilation and lung perfusion scintigraphy used without a carrier as a complement to the ^{99m}Tc-MAA perfusion test. Finally, after describing the clearance mechanism of the ^{81m}Kr tracer, the physico-chemical and pharmacological properties of [^{123}I] MIBG and [^{123}I] Ioflupane radiopharmaceuticals used in SPECT imaging for the functional study of the thyroid gland are discussed. The usefulness of iodine, the production chain of the radiotracer ^{123}l via generators ^{127}I $(p, 5n)$ ^{123}Xe, ^{121}Sb $(\alpha, 2n)$ ^{123}I, ^{122}Te (d, n) ^{123}I and ^{122}Te $(\alpha, 3n)$ ^{123}Xe, the main emissions of the radiotracer ^{123}I, the principle of iodine-123 scintigraphy and its excretion from the body are presented. The physico-chemical and pharmacological properties of radioiodine-131 used in radiotherapy for the treatment of thyroid cancer or metastases, hyperthyroidism and goiter are then studied. This study begins with a description of the ^{131}I radiotracer production chain via the nuclear reactions ^{130}Te (n, γ) ^{131m}Te and ^{130}Te (n, γ) ^{131g}Te, the main emissions of iodine-131, the principle of iodine-131 scintigraphy and its excretion mechanism. Then, the concepts of aerosol therapy, nebulization systems, aerosol generators and median mass aerodynamic diameter (MMAD) are defined. These definitions are rounded off with a description of aerosol deposition phenomena according to their MMAD. This is followed by a presentation of the prostate in the urinary tract and prostate diseases such as adenoma, prostatitis and cancer. These conditions are studied in detail, presenting their causes, risk factors, symptoms, diagnosis and treatment. The main treatment modalities for prostate cancer are then described. The special case of prostate brachytherapy using iodine-125 implants is examined in detail. The study begins with a description of the radioiodine-125 decay diagram, the brachytherapy modes of low-dose-rate (LDR), high-dose-rate (HDR) and pulsed-dose-rate (PDR) brachytherapy, and the procedure for brachytherapy with iodine-125 implants. Prostate chemotherapy is then briefly described.

Chapter 3 concludes with two appendices, shown in sections 3.8.1 and 3.8.2. The first appendix provides general information on strokes. This appendix provides a link between brain scintigraphy with HMPAO or ECD (discussed in section 2.5.3), which is useful for diagnosing cognitive disorders, strokes and epilepsy. In addition, this appendix is very useful for the reader, given that strokes are now known to be very frequent and constitute the leading cause of death in women and the second in men worldwide, and that 30% of stroke victims suffer from depression, according to estimates as of 2019. The second appendix is devoted to general information on thyroid and parathyroid scans, with particular reference to thyroid scintigraphy using radioiodine-123, the properties of which are studied throughout Chapter 3. This appendix is essential for understanding how the thyroid functions and the mechanism of thyroid hormone secretion that determines the various forms of hyperthyroidism (Graves' disease, basedowified goiter, hot nodules, toxic nodules, etc.). Chapter 3 is followed by a list of references, which enables the reader to deepen their knowledge surrounding the issues presented in the book. The book is completed by an index in the final section.

We would like to express our gratitude to Dr. Patrick Berthault, Université de Paris Saclay, CEA Saclay, 91191 Gif-sur-Yvette, France for his valuable review notes on NMR. We would also like to thank Dr Maimouna Mané Cheffe, Head of the Oncology-Radiotherapy Department, Centre Hospitalier National (CHN), and Cheikh Ahmadoul Khadim, Touba, Senegal, for her valuable comments on X-ray imaging.

This book is written for students, teachers and researchers working in the fields of nuclear physics applied to medical imaging in general and nuclear medicine imaging in particular. The book is written in clear, concise language, underpinned by a highly original pedagogical style. As with the first two volumes, each chapter begins with a presentation of the general objective, the specific objectives targeted and the prerequisites necessary for understanding the chapter to be studied. In addition, corrected application exercises and a vocabulary corner defining the key concepts introduced in each paragraph are presented at various points throughout the book. This enables the reader to gain a clear understanding of the knowledge being taught, without the need for a dictionary or glossary.

This book does not attempt to cover every aspect of the description of medical imaging methods and the use of radiopharmaceuticals in nuclear medicine. However, it does provide essential information for readers working in the field of medical imaging. As human work can be perfected, we are always open to suggestions, comments or criticism from our readers that could help improve the scientific quality of this book.

Ibrahima SAKHO
Dakar
April 2024

1

Radiological Imaging

General objective	
Describe the principle of radiological imaging methods.	
Specific objectives	
Define medical imaging;	Describe closed-field MRI and open-field MRI;
Distinguish between radiological imaging and nuclear medicine imaging;	Explain the concept of spin–lattice relaxation;
Know the different medical imaging modalities;	Make the connection between longitudinal magnetization regrowth and T_1 relaxation time;
Know the nature of the radiation or fields used in medical imaging techniques;	Explain the concept of spin–spin relaxation;
Define the linear absorption coefficient;	Relate transverse magnetization decay to T_2 relaxation time after stopping a 90° pulse;
Define the intensity I (x) of a photon beam;	Interpret the evolution with time of the longitudinal and transverse components of the macroscopic magnetization vector after excitation by a 90° RF wave;
Know how the attenuation coefficient changes with atomic number Z and photon energy;	Define the notion of free induction decay (FID);
Express the relationship between the total attenuation coefficient and the coefficients of photoelectric absorption, Compton effect and pair creation;	Distinguish between spin echo and gradient echo sequences;
Define the notion of voxel;	Distinguish between echo time TE and repetition time TR;
Define the Hounsfield unit scale;	Interpret spin dephasing and rephasing phenomena during spin echo signal evolution;

Understand the usefulness of the treatment planning system (TPS) in radiotherapy;	Understand the benefits of applying a 180° pulse to overcome the inherent inhomogeneities of the instrumental magnetic field;
Describe the principle of X-ray production;	Know the intrinsic parameters of tissues that enable their characteristics to be highlighted in MRI;
Describe the principle of X-ray imaging;	Explain the principle of T_1 and T_2 image weighting and proton density;
Describe the principle of computed tomography (CT) or scanner;	Understand the influence of repetition time and echo time on the contrast of MRI images;
Explain the nuclear magnetic resonance (NMR) phenomenon;	Explain the action of MRI contrast agents to improve tissue characterization; Understand the main advantages of the gradient echo sequence;
Define the nuclear magnetic moment;	Define the notion of slice plane;
Describe the physical model of NMR in biological tissues;	Explain the usefulness of gradient slice selection;
Define the Larmor frequency in NMR;	Explain the usefulness of applying a frequency-coding gradient;
Know the order of magnitude of the resonance frequencies of the nuclei involved in NMR;	Define an NMR imaging sequence;
Know the influence of magnetic field intensity on the value of the resonance frequency for a given nucleus;	Explain the principle of eliminating residual magnetization by applying additional gradients at the end of each gradient echo cycle;
Describe the quantum model of Larmor precession;	Define the Fourier plane or k-space;
Explain the mechanism of NMR signal generation at the tissue level;	Illustrate the principle of extracting individual frequencies from a signal by applying the Fourier transform;
Define the macroscopic magnetization vector;	Understand the principle of three-dimensional image acquisition (3DFT) in MRI;
Explain the principle of tissue excitation by a radiofrequency electromagnetic wave;	Explain the principle of three-dimensional imaging using a triple Fourier transform (3DFT);
Describe the quantum model of the magnetic resonance phenomenon;	Explain the principle of ultrasound;
Distinguish T_1 and T_2 NMR relaxation times between healthy and tumor tissues;	Understand the different elements of an ultrasound scan;
Know the two magnetic resonance imaging techniques;	Explain the principle of Doppler ultrasound.
Define magnetic resonance imaging (MRI);	

Prerequisites	
Gibbs or Boltzmann–Gibbs distribution;	Radioactive decay law;
Larmor precession phenomenon;	Evolution operator;
Properties of the proton;	Zeeman Hamiltonian;
Spin magnetic moment;	Properties of ultrasonic waves;
State vector in quantum mechanics;	Doppler effect.

1.1. Radiological imaging and nuclear medicine imaging

1.1.1. *Medical imaging methods*

Medical imaging denotes all means of acquiring and reproducing images of the human body. It is used in two fundamental areas of medicine: radiology and nuclear medicine [BUV 09, ZIM 06, ANS 16, MUL 16]. The various medical imaging methods are based on different physical principles [MUL 16]:

– X-rays: radiography and computed tomography (CT);

– ultrasound;

– nuclear magnetic resonance: MRI;

– radiotracers: scintigraphy and PET (nuclear medicine).

Different types of images can be obtained by varying the type of energy used and the acquisition technology. The various ways of producing medical images are referred to as imaging modalities. Each modality has its own application in medicine [GOU 10]. The following are the different medical imaging modalities:

– transmission imaging: X-ray and computed tomography (CT) or CT scan (X-ray transmission);

– emission imaging: scintigraphy or SPECT (single-photon emission tomography (gamma photon emission)), PET (positron emission tomography), MRI (emission of radiofrequency waves in a static magnetic field);

– reflection imaging: ultrasound (ultrasonic wave reflection).

Figure 1.1 shows the different types of radiation or fields used in radiology and nuclear medicine imaging techniques.

This chapter focuses on radiological imaging methods.

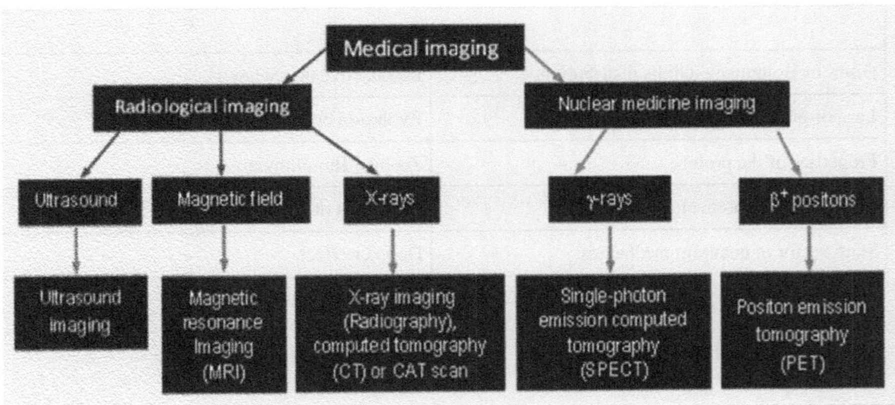

Figure 1.1. *Radiation, fields and imaging techniques in radiology and nuclear medicine*

1.1.2. *Absorption coefficients and Hounsfield number*

Let us consider an electromagnetic beam passing through a homogeneous medium on a surface S. We denote by σ_e the *effective cross-section* (collision surface) of an electron and by n_e the *electron density* (number of electrons per unit volume of the medium traversed). A cylindrical beam volume with axis of revolution Ox therefore contains $n_e \cdot S$ electrons per unit length. By definition, the *total effective cross-section* $\sigma_t = n_e \cdot S \cdot \sigma_e$. During propagation, the beam undergoes attenuation due to absorption by the *absorbing medium*. By definition, the *linear absorption coefficient* μ is given by the relation [DIS 14]:

$$\mu = \frac{\sigma_t}{S} = n_e \cdot \sigma_e \qquad [1.1]$$

Let us denote by N_0 the initial number of photons in the electromagnetic radiation, by N the number of photons remaining (unabsorbed) after attenuation of the beam and by n the number of photons attenuated over a penetration length x of the beam. The fraction dn of photons attenuated over an elementary path dx is then written as:

$$\frac{dn}{n} = -\mu\, dx \qquad [1.2]$$

In equation [1.2], the sign "–" reflects the reduction in the number of photons as the beam propagates through the absorbing medium under consideration.

Integrating [1.2] between the limits N_0 and N gives the number N of photons remaining. Therefore:

$$N(x) = N_0\, e^{-\mu x} \qquad [1.3]$$

In the case of a *heterogeneous medium* (e.g. the human body), we need to take into account the linear absorption coefficient $\mu\,(x)$ characteristic of the medium traversed (e.g. an organ) at the point of abscissa x. For an elementary path dx, equation [1.3] becomes:

$$N(x) = e^{-\int_0^x \mu(x)\,dx} \qquad [1.4]$$

By definition, the *beam intensity* $I\,(x)$ measures the energy of the radiation, or the number of photons per unit time passing through the unit area normal to the beam, at position x. According to [1.3], the intensity $I\,(x)$ of electromagnetic radiation decreases exponentially with the thickness x of the material traversed, according to the law:

$$I(x) = I_0\, e^{-\mu x} \qquad [1.5]$$

The attenuation coefficient μ increases as a function of the atomic number Z of the atoms in the material and decreases as a function of photon energy (in this case, *X photons*). Moreover, this coefficient depends on the chemical composition of the tissues traversed. It is high for bone, medium for soft tissue and low for fat. Bone contains mineral salts (phosphorus, calcium, magnesium), which have higher atomic numbers than the main constituents of soft tissue (oxygen, carbon, hydrogen, nitrogen, etc.). They therefore absorb more *X-rays* [SAK 17].

Taking into account the heterogeneity of the biological media (bone, soft tissue, etc.) through which X-rays pass, we introduce the attenuation coefficient $\mu\,(x)$ specific to each medium. The *total attenuation coefficient* μ is the sum of three coefficients: the *photoelectric absorption coefficient* (τ); the *Compton effect coefficient* (σ); and the *pair creation coefficient* (κ). Therefore:

$$\mu = \tau + \sigma + \kappa \qquad [1.6a]$$

For X-rays, only the photoelectric absorption coefficient and the Compton effect coefficient are taken into account. Pair formation is a specific property of the gamma photon.

For a heterogeneous medium, we deduce the intensity $I(x)$ from equation [1.5]:

$$I(x) = I_0 e^{-\int \mu(x)dx} \quad [1.6b]$$

In *radiography* or *computed tomography (CT)*, an *image* is produced by projecting a beam of X-rays through the object to be imaged. A portion of the X-ray photons is attenuated in the object, and the quantity of photons passing through it is measured by a *detector* located opposite the source. Different interactions are responsible for the attenuation of a photon beam. These interactions depend on two parameters of the material: its electronic cross-section and its electron density.

CT (*Computed Tomography*) images reflect the attenuation coefficient, which depends on electron density and effective cross-section. These two parameters are fundamental quantities for *Monte Carlo simulations* (see note at end of section). A CT device projects at different angles around an object. Upon exiting the object, the beam detected by the CT device contains all ineffective photons (those that have not interacted with the medium or have scattered to reach the device with a deviated trajectory). The CT image can be reconstructed from these projections by filtered back-projection [DIS 14].

In 3D medical imaging (scanner, NMR, PET), each elementary volume of the image is called a *voxel*. The latter is particularly well suited to the representation of a body volume made up of cross-sectional images of a patient. The cross-sectional image of an object is reconstructed from a large number of attenuation measurements at various incidences. A computer then determines the value of the attenuation coefficient μ for each voxel. The Hounsfield unit scale is a linear transformation used to determine the absorption coefficient μ in a given voxel, expressed in *Hounsfield units* (HU). The *average attenuation coefficient* μ for water is given by the following relationship [THI 07, DIS 14]:

$$HU = 1{,}000 \times \frac{\mu - \mu_e}{\mu_e} \quad [1.7]$$

In equation [1.7], μ_e is the *linear attenuation coefficient* of water.

The relationship [1.7] is translated into an experimental curve. This curve is very important for systems calculating the electron density distribution of each voxel, in order to accurately calculate the delivered dose. To perform dose calculations, the value of the attenuation coefficient expressed in HU is first generally converted into electron density. The latter is typically obtained by means of a stoichiometric calibration using images of a calibration phantom. This phantom contains a series of 13 inserts in a uniform medium. The HU numbers obtained in the image are compared with the density values of the insertions. A two-, three- or four-line regression is performed on the measurements obtained from the phantom, to form the HU–ED (Hounsfield Unit–Electron Density) curve. These lines are defined using a Matlab1 procedure developed for clinical use. This curve is then used to convert HU numbers from clinical images into electron density [DIS 14].

NOTE.–

Godfrey Newbold Hounsfield (1919–2004) was a British engineer, famous for having designed the first scanner in 1971. He built a computer that takes X-ray images of the same object from different angles, reconstructing an image of the object in slices. The application of this technique to the medical field led him to propose what we now call tomodensitometry or CT scanning. The mathematical theory behind the scanner was independently developed by South African physicist Allan MacLeod Cormack (1924–1998), who became a naturalized American citizen in 1966. Hounsfield and Cormack each received half of the 1979 Nobel Prize in Physiology and Medicine for the development of computerized tomography, known as CT scanning. The Hounsfield scale describing *radiodensity* (*opacity of a material to electromagnetic radiation in general and X-rays in particular*) and the Hounsfield unit are named in his honor.

1.1.3. *Treatment planning system (TPS)*

In *radiotherapy*, treatment preparation is carried out through a treatment planning system (TPS) to determine the *dose distribution* in the *tumor* and surrounding areas. This dose distribution must be optimized so that healthy tissue around the tumor receives the minimum dose. Many of the TPSs in use incorporate dose distribution calculation algorithms, medical image processing and tumor contouring tools.

Three fundamental points are necessary for accurate planning:

1) the use of high-quality scanned images that reliably represent the patient in the treatment position, allowing the identification of tumor structures and organs at risk. This type of imaging, thanks to image registration techniques, can be combined with other modalities such as MRI or PET;

2) precise simulation of the interaction of ionizing radiation with human tissue;

3) *dose calculation* based on the physical and geometric parameters of the treatment device and the patient.

In general, any dose planning system must satisfy a number of constraints:

– a good description of the physical processes involved in particle generation;

– patient-related heterogeneity correction with primary and secondary particle transport;

– a dose calculation at a possible random point before calculating the 3D matrix;

– on-axis dose calculation accuracy criteria for a calculation resolution of at least 5 mm $<$ 1% and off-axis $<$ 2%.

The quality and precision of this simulation phase is obviously linked to the accuracy of the dose calculation models and algorithms. Two main difficulties are generally encountered in dosimetric calculations with TPS: firstly, modeling the patient (a living being) with a tissue architecture that is highly complex and even virtually impossible to model completely, and secondly, treatment constraints. It is therefore necessary and essential to make approximations and optimum compromises between calculation accuracy and calculation speed.

For modeling purposes, the patient is considered as a volume made up of a set of small elements (voxels) obtained from the CT scan. As mentioned above, each voxel contains an H (Hounsfield) number representing the *attenuation of X-rays* in this volume element, and providing access to electron densities for dosimetric calculations within the voxel. Dosimetry involves calculating the physical interaction of particles with this voxel volume.

Once the patient has been modeled, the dose is calculated using models that take into account treatment parameters such as accelerator parameters (energy, collimation, intensity, etc.), as well as anatomical data. TPS currently delivers satisfactory results. However, in order to improve accuracy, and in certain situations where TPS accuracy is not sufficient, *Monte Carlo methods* may be considered. Monte Carlo simulations now represent a powerful alternative and tool for modeling dose response and calculating the corrections required for dosimetric applications. The use of these methods for complex geometries or high-resolution images requires relatively long computing times, which must be taken into account.

NOTE.–

Monte Carlo methods [ROG 82, THI 07, SPA 12, DIS 14, PET 14, DEG 17] are statistical methods based on the drawing of pseudo-random numbers, according to probability laws or probability density functions that describe natural processes (biological or physical). The name of these methods, which alludes to the games of chance played at the Monte Carlo casino (in the municipality of Monaco), was invented in 1947 by the Greek-American physicist Nicholas Constantine Metropolis (1915–1999). It was first published in 1949 in an article coauthored with the Polish mathematician Stanisław Marcin Ulam (1909–1984). In the field of medical physics, particularly for dosimetric applications in external radiotherapy/curietherapy, Monte Carlo methods use various types of software resulting from collaborations between medical physicists and computer scientists. These methods are underpinned by several calculation codes, each specific to a given application. Examples include the *PENELOPE* (*PENetration and Energy Loss of Positrons and Electrons*) code [BAR 95] for the simulation of electron–photon cascades. This simulation is carried out in an energy range from 1 keV (100 eV for electrons and positrons) to a few hundred MeV in base materials with atomic numbers between 1 (hydrogen) and 98 (californium), as well as in compound materials. In addition, the *MCNP (Monte Carlo Neutron Photon*) code [BRE 00] enables the simulation of photons, neutrons, positrons and electrons. This code enables photons to be "tracked" over a wide range (from 1 keV to several GeV). This code has evolved into its new version MCNP6 following a merger of versions MCNP5 and MCNPX (*Monte Carlo N-Particle eXtended*).

VOCABULARY CORNER.–

Algorithm: the description of a sequence of steps used to obtain a result from input elements.

1.2. X-ray imaging

1.2.1. *Principle of X-ray production*

X-ray imaging uses two main instruments: an X-ray source and an X-ray detector, positioned on either side of the patient. The principle of X-ray production is illustrated in Figure 1.2 [SAK 19]. X-rays are produced in *X-ray tubes*, also known as *Coolidge tubes or hot-cathode tubes.*

Let us briefly review the principle of X-ray production.

Electrons emitted by a cathode (*a filament, usually made of tungsten, heated by the passage of an electric current*) are accelerated by a high potential difference (from 10 to 150 kV) towards a target made of a metal anode (also made of tungsten). X-rays are emitted by the target in two ways:

– The stopping of the electrons by the atoms of the target creates *continuous radiation* (*braking radiation or Bremsstrahlung*), part of which is in the X-ray range.

– The accelerated electrons have sufficient energy to excite some of the target atoms, disrupting their inner electronic layers. These excited atoms then emit X-rays as they return to their ground state.

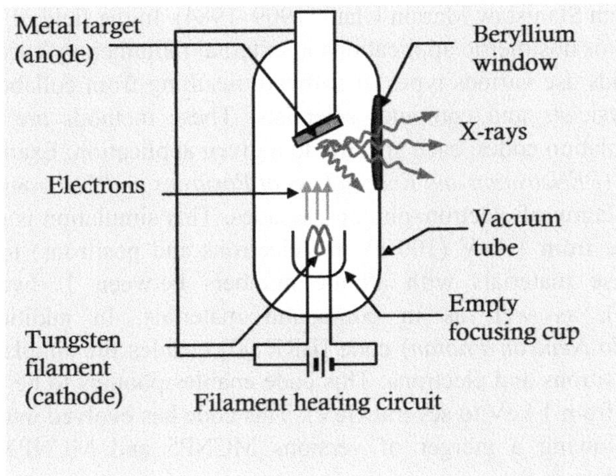

Figure 1.2. *Principle of X-ray production using a Coolidge tube. For a color version of this figure, see www.iste.co.uk/sakho/nuclearphysics3.zip*

1.2.2. *Principle of X-ray imaging*

The oldest noninvasive medical imaging technique applied to the skeleton is X-ray radiology. This early technology was born after the discovery of X-rays in 1895 by Wilhelm Roentgen, who also made the first images of human anatomy. Roentgen experimented with X-ray imaging by placing his wife Anna Bertha Ludwig's hand in the path of the X-rays. He found that the image obtained was a shadow of the hand bones, with her wedding ring visible (Figure 1.3). The bones are surrounded by a penumbra representing the flesh of the hand. The flesh is therefore more permeable to X-rays.

Roentgen's first X-ray therefore gave birth to the discipline known as radiology. Radiography is based on two main pieces of equipment: an X-ray source positioned in front of the patient and an X-ray detector (usually flat) placed on the other side (Figure 1.4).

The principle of this technique is based on the attenuation properties of X-rays by solid, liquid or gaseous materials exposed to X-ray radiation. The absorption of this radiation by the various materials through which it passes enables the formation of an X-ray image on a screen or detector.

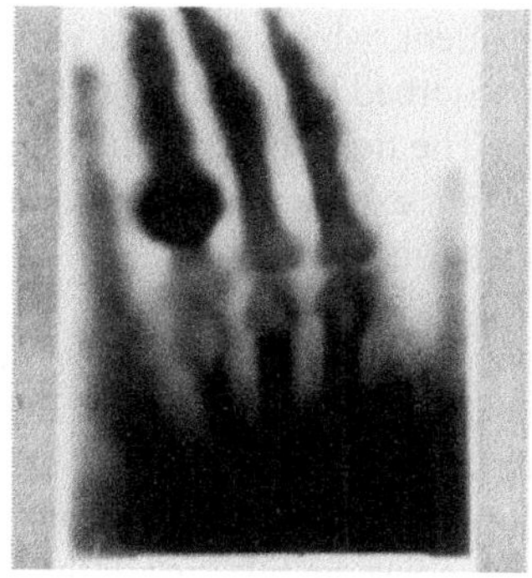

Figure 1.3. *First X-ray image of Anna Bertha Ludwig's hand. Image taken on December 22, 1895 (source: https://en.wikipedia.org/wiki/Wilhelm_R%C3%B6ntgen)*

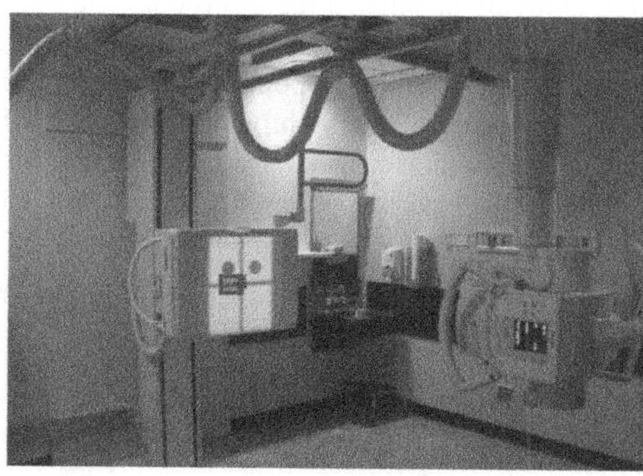

Figure 1.4. *X-ray equipment [ECO 13]. For a color version of this figure, see www.iste.co.uk/sakho/nuclearphysics3.zip*

Since the 1950s, X-ray imaging technology has undergone a revolution with the advent of digital technology. Unlike conventional radiology, *digital radiology* differs in its acquisition chain, which comprises a *luminance amplifier* that transforms X-rays into visible photons, and a coupling optic linked to a high-performance video camera. The video signal is digitized, visualized and stored in computer memory. It should be noted, however, that three-dimensional use of X-ray images is difficult: organs are superimposed without any indication of position other than their projection on the screen [DIL 92].

During an *X-ray examination*, X-rays are transmitted to the patient. The basic process consists of a short-duration (0.5 s) emission of X-rays from the source positioned in front of the patient, which interacts with the patient [ECO 13]. These rays are then more or less absorbed by the biological media they pass through, with an absorption coefficient μ. The detector used then highlights the attenuation μ (x), which varies according to an exponential law taking into account photoelectric absorption and Compton scattering.

The initial homogeneous distribution of rays (those leaving the source) is modified according to the intensity with which they are absorbed (attenuated) or scattered in the body. The attenuation properties of tissues such as bone or soft tissue are different, resulting in a nonhomogeneous distribution of the rays that emerge from the patient and, consequently, reach the detector plate. In this way, the final radiographic image obtained corresponds to the image of the distribution of attenuations due to the various organs successively passed through. The detector can be either a photosensitive film or an electronic detection system (digital radiography) [ECO 13]. Figure 1.5 shows the principle of radiology acquisition. It shows white areas corresponding to highly attenuated regions (bone) and black areas (soft tissue) reflecting less attenuated areas. As the X-ray source is outside the body, this is called *transmission imaging*, where each point in the image corresponds to a piece of information along a linear trajectory through the patient's body. This type of imaging is widely used for diagnosing bone fractures, lung cancer or cardiovascular problems [ÉCO 13, POI 19].

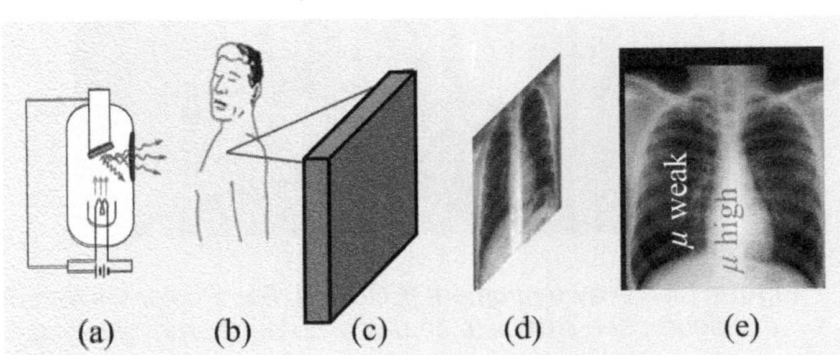

Figure 1.5. *Principle of X-ray acquisition. (a) X-ray tube, (b) patient, (c) detector + reader (conversion of X-ray photons into an image), (d) projection (of the attenuation coefficients μ of the structures crossed by the X-rays), (e) grayscale image (corresponds to the sum of the attenuations μ of the organs successively crossed. Objects appear superimposed on one another). For a color version of this figure, see www.iste.co.uk/sakho/nuclearphysics3.zip*

X-rays are also ionizing radiation, so X-ray imaging is said to be ionizing. During a radiological examination, the patient receives an equivalent dose of radiation, measured in *sieverts*. The purpose of this dose is to measure the impact of exposure to a radioactive source on biological tissues and to highlight any radiation-related danger. For this reason, it is not advisable to have more than one examination every three months. To reduce radiation exposure during an examination, new and increasingly sensitive detectors have been developed. Film detectors (initially silver film) have gradually given way to digital detectors and now to planar sensors [POI 19].

As shown in the preceding X-ray image (Figure 1.4), X-ray imaging enables us to differentiate between calcific structures (bone), soft tissues (flesh), air and gases in the human body. In the case of *contrast radiography,* often used in gastroenterology, it may be necessary to use a contrast medium that is highly opaque to X-rays. This product provides an X-ray image of the organ in which it is placed. This procedure was used to produce the first blood circulation images, as well as images of liver, stomach and bile functions. Radiology, limited to *anatomical imaging*, is used in various fields of medicine, such as orthopedics, rheumatology, orthodontics, pneumology (preventive radiography of the lungs, suspicion of lung cancer) and oncology (mammography, breast cancer screening) [ZIM 06].

VOCABULARY CORNER.–

– *Gastroenterology*: a medical specialty that studies the digestive system and its diseases.

– *Luminance*: a quantity corresponding to the visual sensation of brightness of a surface. It is defined as the power of visible light passing through or being emitted from a surface element in a given direction, per unit area and per unit solid angle (see Volume 1). Image luminance is measured in candela per square meter (cd/m^2). In photo retouching software, the term is used to designate the brightness of a color. The brighter a hue, the more white it contains. In electronics, a brilliance or luminance amplifier is a system that enables the detection of low light intensities, or even the transformation of an optical image into an electronic image. The purpose of the brightness amplifier is to increase the luminosity of the images thus obtained, and above all their precision in radiology.

– *Mammography*: an X-ray of the breasts to detect abnormalities.

– *Orthodontics*: a dental specialty dedicated to correcting malpositioned jaws and teeth to optimize the postural balance between bone structures and occlusion, as well as the development of bone bases for functional and esthetic purposes.

– *Orthopedics*: a surgical specialty concerned with the prevention and correction of disorders of the musculoskeletal system, including deformities and malformations of bones, joints, muscles, tendons and nerves.

– *Oncology*: a medical specialty concerned with cancerous tumors.

– *Pneumology*: a specialty concerned with the functioning of the respiratory system – lungs, bronchi, pleura and trachea – and its diseases.

– *Rheumatology*: a medical specialty dedicated to the treatment of diseases of the musculoskeletal system: in particular, diseases of the bones, joints and muscles. Rehabilitation is a crucial aspect of the treatment of these diseases.

– *Biological tissue*: a collection of similar cells of the same origin, grouped together in a cluster, network or bundle (fiber). A tissue forms a functional whole, i.e. its cells work together to perform the same function. Biological tissues regularly regenerate and are assembled together to form organs (heart, lungs, etc.). The science that studies tissues is histology.

NOTE.–

In medical imaging, a *contrast agent* is a substance that artificially enhances contrast to visualize either an anatomical structure such as an organ or a pathological structure such as a tumor. It is used when these structures have little or no natural contrast (i.e. it is difficult to distinguish them from neighboring tissues). The contrast medium used in contrast radiography is a nontoxic liquid, often based on iodine or barium [CEN 21c]. The action of the contrast agent administered depends on the medical imaging technique used (in radiography, the agent's ability to absorb X-rays is exploited; in MRI, the compounds used are chosen on the basis of their magnetic properties; in ultrasound, substances with a characteristic ultrasound echo are used, etc.).

Wilhelm Roentgen (1845–1923) was a German physicist. He is best known for his discovery of X-rays, which paved the way for medical X-ray imaging in radiology. He was awarded the first Nobel Prize in Physics in 1901 for this important discovery.

1.2.3. *Principle of computed tomography or scanner*

In addition to radiology, computed tomography (CT) or scanner is another medical imaging technique that involves measuring the absorption of X-rays by tissues and then, through computer processing, digitizing and finally reconstructing 2D or 3D images of anatomical structures. There are several names for this technique: scan, computed axial tomography (CAT), CT-scan (computed tomography-scan), CAT-scan (computer-assisted tomography-scan) or simply scanner. Data are acquired using tomographic or "slice" analysis, by scanning the patient with a beam of X-rays.

CT provides good distinction between bone and soft tissue. Contrast agents can be used to improve soft-tissue differentiation. Although magnetic resonance imaging is gradually replacing X-ray scanning for volume exploration, the latter remains one of the main tools for 3D acquisition, due to its widespread use in hospitals, its lower cost than MRI and its good bone detection [DIL 92].

In Figure 1.6, the green beams are positioning laser beams used to define the *scanner zero*. This zero defines the origin of a three-dimensional reference frame in which the coordinates (X, Y, Z) of the barycenter of the tumor to be diagnosed are defined. This marker is marked by three radio-opaque beads placed on the patient. These markers are then "tattooed" and used to reposition the patient prior to treatment.

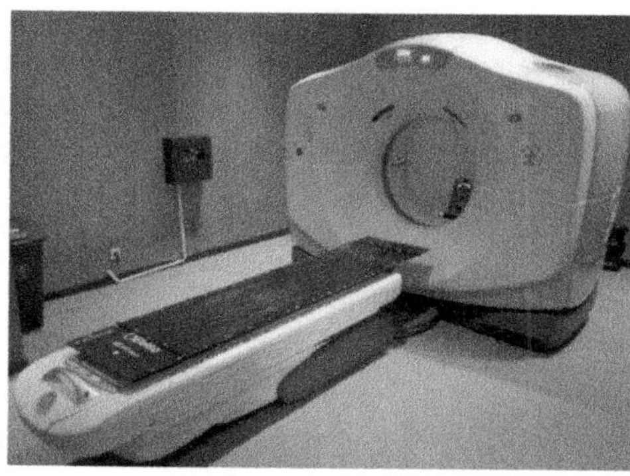

Figure 1.6. *Scanner (dedicated to radiotherapy) installed at the Centre International de Cancérologie de Dakar (CICD). For a color version of this figure, see www.iste.co.uk/sakho/nuclearphysics3.zip*

A CT scan is not only extremely useful for tumor location, but also indispensable for dosimetry in radiotherapy. It provides the information needed to simulate the deposition of radiation energy in the patient [THI 07]. Therefore, prior to radiotherapy, the CT scanner is used to obtain an image of the tumor to be treated.

The X-ray tube used in CT (Figure 1.7) consists of a cathode emitting electrons that produce X-rays (between 80 and 150 kV) when they strike the tube's anode.

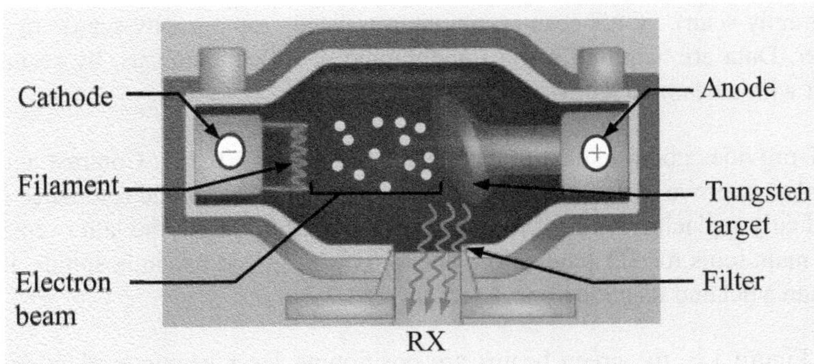

Figure 1.7. *X-ray tube used in CT. For a color version of this figure, see www.iste.co.uk/sakho/nuclearphysics3.zip*

The principle of CT scanning is based on the use of the X-ray absorption coefficients of the human body [THI 07]. During a scanning session, the patient lies down and is injected with a contrast medium to improve the quality of the image to be obtained. The X-ray source and detector rotate around the patient (Figure 1.8). The X-rays generated in the X-ray tube pass through the patient's body, interacting to varying degrees with the tissues.

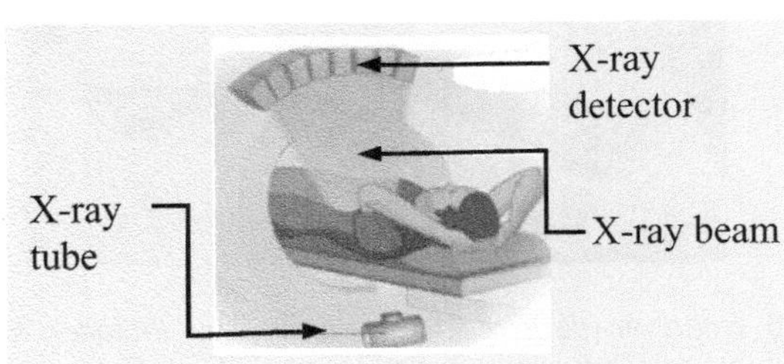

Figure 1.8. *Patient position during an X-ray scanning session. For a color version of this figure, see www.iste.co.uk/sakho/nuclearphysics3.zip*

As the tube–detector assembly rotates around the patient, a computer stores the intensity of the X-ray beam received by the detector in the form of successive "snapshots". This digitized data is then used to calculate elementary views (Figure 1.9).

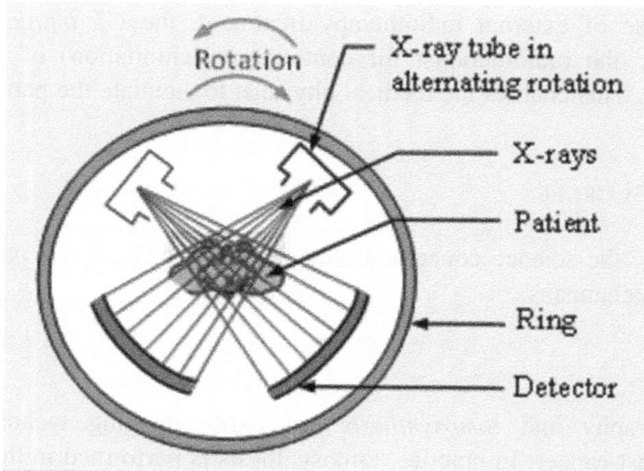

Figure 1.9. *Alternating rotation of the tube–detector assembly around the patient for imaging. For a color version of this figure, see www.iste.co.uk/sakho/nuclearphysics3.zip*

Each image displayed corresponds to the shadow cast on the irradiated organs. By recording a succession of light and dark zones, the computer deduces an elementary image, which it memorizes along with its orientation at the time of acquisition. The computer then combines all of the elementary images to create the final image. Using a digital calculation system, it rotates each elementary image to restore its original orientation. Finally, it "superimposes" them to reconstruct the image of each organ displayed on the screen (Figure 1.10). This enables the radiologist to highlight the pathology in question.

Figure 1.10. *Computer processing of elementary images to reconstruct the final image of each organ through which X-rays pass (here Centre International de Cancérologie de Dakar). For a color version of this figure, see www.iste.co.uk/sakho/nuclearphysics3.zip*

In the case of external radiotherapy treatment, the *CT images* obtained are transmitted to the radiotherapist for contouring (delimitation) of the tumor and organs at risk. This enables the medical physicist to simulate the patient's treatment plan.

VOCABULARY CORNER.–

Pathology: the science concerned with the study of diseases, particularly their causes and mechanisms.

NOTE.–

Mammography and *tomosynthesis* are X-ray imaging techniques used to diagnose breast cancer. In practice, tomosynthesis is performed in the same way as mammography. What all tomosynthesis systems have in common is that the X-ray tube moves over a limited arc above the compressed breast, enabling projection images to be obtained at multiple angles of incidence. The tube moves either continuously or step by step. The detector located directly under the breast remains fixed for most manufacturers, but can be mobile depending on the X-ray photon detection technology used [DÜR 15].

1.3. Nuclear magnetic resonance (NMR)

1.3.1. *Definition and principle of NMR*

Nuclear magnetic resonance (NMR) is an experimental method for studying the spins of atomic nuclei. The principle is as follows. Consider a material with a magnetic moment $\vec{\mu}$ placed in an *isotropic medium*. By applying a *static magnetic field* $\vec{H}$, the moment $\vec{\mu}$ orients itself in the direction of $\vec{H}$ and begins to execute a Larmor precession motion around it at the *Larmor frequency* $\omega_0 = \gamma H$, where γ denotes the gyromagnetic ratio. Next, let us apply to the material a radiofrequency field of frequency ω_0, perpendicular to the direction of the field $\vec{H}$. The material then enters resonance, absorbing electromagnetic energy. This is the *phenomenon of magnetic resonance*. When the magnetic moment arises from the electron spin, we speak of electron paramagnetic resonance. When the magnetic moment arises from nuclear spin, the excitation field is in the microwave range, and this is known as nuclear magnetic resonance.

Chronologically speaking, the phenomenon of magnetic resonance was demonstrated in 1946 by Edward Mills Purcell and his colleagues [PUR 46] at Harvard University, independently of Felix Bloch [BLO 46] at Stanford University.

1.3.2. *Nuclear magnetic moment*

As a general rule, any charged particle describing a closed curve is a source of magnetic field. At a great distance from the particle, the magnetic field can be considered as that created by a magnetic dipole placed at the location of the charged particle [SAK 18].

As with electrons, the nucleus has a *spin angular momentum* $\vec{I}$, which is an observable. The square $\vec{I}^2$ and one of its projections, for example I_z along a preferred direction (here Oz), have determined values equal to $\hbar^2 I(I+1)$ and $M_I \hbar$ respectively. The *nuclear magnetic quantum number* M_I then takes on the values $-I$, $-(I+1)$, ….., $+(I-1)$, $+I$, i.e. a total of $(2I+1)$ values.

By definition, the *nuclear dipole magnetic moment* $\overrightarrow{M}_N$ is given by the relation [SAK 20a]:

$$\overrightarrow{M}_N = g_I \gamma_I \vec{I} \qquad [1.8]$$

In relation [1.1], g_I is the *nuclear Landé factor* and the quantity γ_I is the *nuclear gyromagnetic ratio* given by the following expression (it is the analogue of the *electron's gyromagnetic ratio* $\gamma = -e/2m$):

$$\gamma_I = \frac{e}{2m_p} \qquad [1.9]$$

In this relationship, m_p is the mass of the proton.

Similarly, we define the *nuclear magneton*, also known as the *Bohr nuclear magneton*, noted μ_N (by analogy with the *Bohr magneton* $\mu_B = -\gamma\hbar = e\hbar/2m$).

The nuclear magneton is given by the expression:

$$\mu_N = \gamma_I \hbar = \frac{e\hbar}{2m_p} \qquad [1.10]$$

Note that $\mu_N/\mu_B = m/m_p$ and $m_p/m = 1{,}836.1526715$. Furthermore, numerically, the Bohr magneton $\mu_B = 9.27400915 \times 10^{-24}$ J · T^{-1}, so:

$\mu_N = 5.05078317 \times 10^{-27}$ J · T^{-1}.

Using [1.10], the nuclear magnetic moment [1.8] can be written as:

$$\vec{M}_N = g_I \frac{\mu_N}{\hbar} \vec{I} \qquad [1.11]$$

For the proton:

$I = 1/2;\ g_I = 5.5883;\ \mu_p = 2.7928473375\ \mu_N.$

Although electrically neutral, the *neutron's spin angular momentum* is associated with a dipole magnetic moment μ_n. For the neutron, this is:

$I = 1/2;\ g_I = -\ 3.8263;\ \mu_n = -\ 1.91304272\ \mu_N.$

When a nucleus with a nonzero spin is subjected to a uniform magnetic field, its magnetic moment interacts with the magnetic field. This results in an additional *magnetic interaction potential energy* given by the relation:

$$W = -\vec{M} \cdot \vec{H}. \qquad [1.12]$$

In relation [1.12], $\vec{M}$ is the sum of the orbital magnetic moment and the spin magnetic moment of a nucleon (or electron in the case of a hydrogenoid system, for example).

1.3.3. *Physical model of NMR in biological tissues*

The main nuclei of biological interest with magnetic properties are the ^1H and ^2H nuclei of the hydrogen atom, the isotopes ^{13}C of carbon, ^{15}N of nitrogen, ^{19}F of fluorine, ^{31}P of phosphorus, etc. Of these nuclei, only the *hydrogen* nucleus (*proton)* plays an important role in imaging (often referred to as "proton MRI"). This is because hydrogen atoms (making up 10% of body mass) account for 86% of the body's chemical composition, and are mainly distributed in water and fat molecules. This makes hydrogen the atom of choice for NMR studies in biological environments [EUS 15]. The proton present in the body has a high intrinsic (protonic) magnetic moment, giving rise to a clear resonance phenomenon. This *intrinsic magnetic moment* is represented by a "microscopic" *magnetization vector* $\vec{\mu}$ aligned with the proton's own axis of rotation. The value of this vector is $\mu = M_N/g_I$. Using [1.8], we obtain:

$$\vec{\mu} = \gamma_I \vec{I} \qquad [1.13]$$

In biological matter, the rotational motion of positively charged water protons induces an internal magnetic field (the proton can be likened to a current-carrying coil). In the absence of an external magnetic field, the distribution of internal magnetic fields is random. Overall, these fields tend to cancel each other out.

In the presence of an intense external static magnetic field $\vec{H}_0$, the proton spins will align themselves parallel or antiparallel to the applied external field. There are two possible scenarios:

– A first half N_+ of spins will orient in the direction of the static field $\vec{H}_0$ (spin up, orientation parallel to $\vec{H}_0$ corresponding to a projection of the spin angular momentum $\vec{I}$: $M_I = + 1/2$).

– Another half N_-, slightly smaller in number, will orient itself in the opposite direction to $\vec{H}_0$ spin down, antiparallel to $\vec{H}_0$, corresponding to a projection of the angular momentum of spin $\vec{I}$: $M_I = - 1/2$).

Consider a set of N protons, each with a spin $I = 1/2$. Let $|+\rangle$ be the microstate of a spin-up proton (+ 1/2) and $|-\rangle$ the microstate of a spin-down proton (– 1/2). The N spins are subjected to the Oz-directed $\vec{H}_0$ field. In this field, the interaction potential energy W of a proton's nuclear magnetic moment $\vec{\mu}$ is given by relation [1.7] ($\vec{M}$ has been replaced by $\vec{\mu}$ in relation [1.5] and $\vec{H} = \vec{H}_0$):

$$W = -\vec{\mu} \cdot \vec{H}_0 = -\gamma_I \vec{I} \cdot \vec{H}_0 \qquad [1.14]$$

Knowing that $\vec{H}_0 = H_0 \vec{z}$ and that the projection of the nuclear angular momentum is quantized ($I_z = M_I \hbar$), equation [1.14] gives:

$$W = -\gamma_I I_z H_0 = -\gamma_I M_I \hbar H_0 \qquad [1.15]$$

For simplicity, let us say $N = 7$ protons. One half of the N_+ spins ($M_I = + 1/2$) are oriented in the direction of the $\vec{H}_0$ field, and the other half of the N_- spins ($M_I = - 1/2$) are oriented in the opposite direction of the static $\vec{H}_0$ field. The application of the magnetic field causes the N spin level to split into two sublevels by the *Zeeman effect*. The energies W_+ and W_- of these sublevels are given by the relations from [1.15]:

$$\begin{cases} W_+ = -\dfrac{\hbar}{2}\gamma_I H_0 \\ W_- = +\dfrac{\hbar}{2}\gamma_I H_0 \end{cases} \qquad [1.16]$$

Relations [1.16] show that the *Zeeman sublevel* of energy W_- is above the Zeeman sublevel of energy W_+ (Figure 1.11). For purely pedagogical purposes, we have arbitrarily set $N_+ = 4$ and $N_- = 3$ (bearing in mind that N_- is slightly lower than N_+).

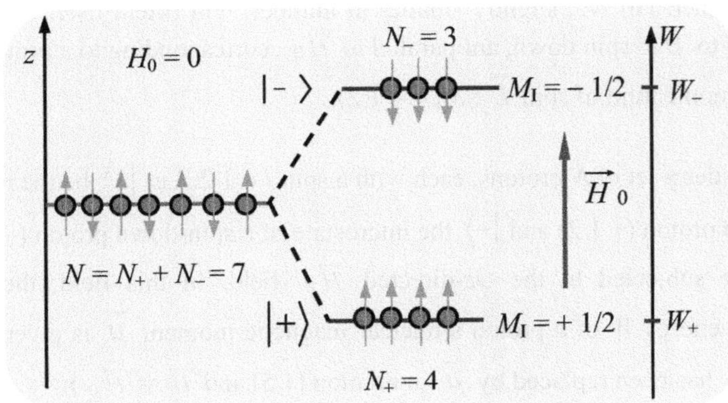

Figure 1.11. *Zeeman sublevels induced by the application of a magnetic field to a set of N = 7 spins. The N₊ population of the lower sublevel (ground level) is slightly higher than the N₋ population of the upper sublevel. Note that this is an illustrative diagram. The number of spins is very high. For a color version of this figure, see www.iste.co.uk/sakho/nuclearphysics3.zip*

Note that the number of spins shown in Figure 1.11 is for educational purposes only. It is not possible to represent the exact number of spins (of the order of few million) for each level. With a total number of $N = 7$ spins, the separation of the N-spin level into two Zeeman sublevels by application of the external magnetic field is clearly illustrated. The value $N = 7$ gives a polarization $(N_+ - N_-)/(N_+ + N_-) = 1/7 = 0.14$. However, at the highest magnetic fields in NMR, this polarization does not exceed 10^{-4}. We will check this result in Application 2.1 with $N = 10^6$ protons.

In the general case, the distribution of the N_+ and N_- populations of the N spins in the magnetic field obeys the Gibbs distribution (also known as the *Boltzmann–Gibbs distribution*). Therefore, within the framework of this distribution, in the equilibrium state at temperature T, the probability P_i of finding a quantum system of energy E_i in the microstate i is given by the relation [SAK 17]:

$$P_i = \frac{e^{-E_i/kT}}{\sum_i e^{-E_i/kT}} \quad [1.17]$$

In [1.17], k is Boltzmann's constant ($k = 1.3806504 \times 10^{-23}$ J · K^{-1}).

Using [1.16], the probabilities $P_\pm$ for a spin ± 1/2 to occupy one of the microstates $|\pm\rangle$ are given by the relations ($E_i = W_i$) (i = + or –):

$$\begin{cases} P_+ = \dfrac{e^{-W_+/kT}}{e^{-W_+/kT} + e^{-W_-/kT}} \\ P_- = \dfrac{e^{-W_-/kT}}{e^{-W_+/kT} + e^{-W_-/kT}} \end{cases} \quad [1.18]$$

In the framework of the *Boltzmann–Gibbs distribution*, the population N_i of a microstate i is given by the relation:

$$N_i = NP_i;\ N = \Sigma N_i \quad [1.19]$$

Using [1.18], we express the populations N_+ and N_- as:

$$\begin{cases} N_+ = N\dfrac{e^{-W_+/kT}}{e^{-W_+/kT} + e^{-W_-/kT}} \\ N_- = N\dfrac{e^{-W_-/kT}}{e^{-W_+/kT} + e^{-W_-/kT}} \end{cases} \quad [1.20]$$

Considering the expressions [1.16], equations [1.20] can be written as:

$$\begin{cases} N_+ = N\dfrac{e^{\hbar\gamma_I H_0/2kT}}{e^{\hbar\gamma_I H_0/2kT} + e^{-\hbar\gamma_I H_0/2kT}} \\ N_- = N\dfrac{e^{-\hbar\gamma_I H_0/2kT}}{e^{\hbar\gamma_I H_0/2kT} + e^{-\hbar\gamma_I H_0/2kT}} \end{cases} \quad [1.21]$$

As shown in Figure 1.11, in the thermodynamic equilibrium state, the lowest-energy state (W_+ energy state) corresponding to the alignment of the spins along the direction of the applied field is statistically favored. We show (see Application 1.1) that, at a temperature of 300 K and for H_0 = 1.5 T, there is a difference of five protons in favor of low-energy protons out of a total of 1 million protons.

APPLICATION 1.1.–

Show that at room temperature (300 K) and for H_0 = 1.5 T, there is a difference of five protons in favor of the set of low-energy protons out of a total of N = 1 million protons. Deduce that the ratio $\Delta N/N = 5 \times 10^{-6}$, $\Delta N = (N_+ - N_-)$. Assume $\varepsilon = \hbar \gamma_I H_0 / 2kT$.

Data: $\gamma_I/2\pi = 42.576$ MHz · T^{-1} [COM 13]; $h = 6.63 \times 10^{-34}$ J · s.

ANSWER.–

Knowing that $\varepsilon = \hbar \gamma_I H_0 / 2kT$, the expressions [1.21] can be written as:

$$N_+ = N\frac{e^{\varepsilon}}{e^{\varepsilon} + e^{-\varepsilon}}; N_- = N\frac{e^{-\varepsilon}}{e^{\varepsilon} + e^{-\varepsilon}} \qquad [1.22]$$

Let us rewrite the ε constant as follows:

$$\varepsilon = h \times \frac{\gamma_I}{2\pi} \times \frac{H_0}{2kT}$$

Numerically, we find:

$$\varepsilon = 6.63 \times 10^{-34} \times 4.2576 \times 10^7 \frac{1.5}{2 \times 300 \times 1.3806504 \times 10^{-23}}$$

$$= 5.1 \times 10^{-6}$$

Using this result, the relations [1.22] give ($N = 10^6$ protons):

$e^{\varepsilon} = 1.000\ 00511$; $e^{-\varepsilon} = 0.999\ 994\ 890$

For the N_+ and N_- populations, we obtain the following values:

N_+ = 500 0002.555 ≈ 500 0002 protons; N_- = 499 9997.445 ≈ 499 9997 protons

We then obtain:

$$\Delta N = (N_+ - N_-) = 5 \text{ protons and polarization } \Delta N/N = 5 \times 10^{-6}$$

By convention, the angular velocity $\Omega_0 = \gamma_I B_0$ expressed in radians per second ($rad \cdot s^{-1}$) is called the Larmor frequency. As shown in equation [1.2], the gyromagnetic ratio γ_I depends on the nucleus under consideration. For all nuclei of interest in NMR, such as ^1H, ^2H, ^{13}C, ^{19}F, ^{31}P and ^{23}Na, the ^1H proton has the highest *resonance frequency* value, as shown in Table 1.1.

Nucleus	Resonance frequency ν_0 (in MHz)
^1H	100.1
^{19}F	94.1
^{31}P	40.5
^{13}C	25.2
^2H	15.4
^{15}N	10.1

Table 1.1. *Resonance frequencies of various nuclei of NMR interest. Values are expressed in MHz, the magnetic field value being equal to 23,500 gauss [HAM 90]. 1 gauss = 10^{-4} T*

APPLICATION 1.2.–

Calculate the resonance frequencies for the magnetic fields (in gauss) obtained in some commercial devices: 14,100, 23,500, 51,700 and 70,500 [HAM 90]. The results will be presented in tabular form (Table 1.2). Conclude. Given: $\gamma_I/2\pi$ = 42.576 $MHz \cdot T^{-1}$ [COM 13].

ANSWER.–

We have:

$$\nu_0 = \Omega_0/2\pi = \gamma_I B_0/2\pi \quad [1.23]$$

For B_0 = 14,100 gauss = 1.41 T, we obtain:

$$\nu_0 = 42.576 \times 1.41 = 60.032 = 60 \text{ MHz.}$$

Considering other values of the magnetic field, equation [1.23] gives the results shown in Table 1.2.

Magnetic fields B_0 (in gauss)	Resonance frequencies ν_0 (in MHz)
14,100	60
23,500	100
51,700	220
70,500	300

Table 1.2. *Proton magnetic resonance frequencies as a function of magnetic fields obtained in some commercial devices [HAM 90]*

Conclusion: the value of the resonance frequency increases with magnetic field strength for a given nucleus.

1.3.4. *Quantum model of Larmor precession*

Consider the relation [1.8]. Replace the nuclear spin I_z by the spin observable S_z. We obtain:

$$W = -\gamma_I S_z H_0 \qquad [1.24]$$

Let us introduce the frequency $\Omega_0 = \gamma_I H_0$. By definition, the *Zeeman Hamiltonian* W_Z associated with the magnetic interaction potential energy W is written as:

$$W_Z = -\Omega_0 S_z \qquad [1.25]$$

In equation [1.25], S_z has eigenvalues $+\hbar/2$ and $-\hbar/2$ associated with eigenkets $|+\rangle$ and $|-\rangle$ respectively. This translates into the eigenvalue equations:

$$\begin{cases} S_z|+\rangle = +\dfrac{\hbar}{2}|+\rangle \\ S_z|-\rangle = -\dfrac{\hbar}{2}|-\rangle \end{cases} \qquad [1.26]$$

Furthermore, relation [1.25] expresses that the observables W_Z and S_z commute and have the same eigenvectors (those of S_z): they therefore constitute a CSCO (Complete Set of Commuting Observables). Using [1.25], the eigenvalue equations of the Zeeman Hamiltonian can be written as follows:

$$\begin{cases} W_Z|+\rangle = S_z|+\rangle = -\frac{\hbar}{2}\Omega_0|+\rangle \\ W_Z|-\rangle = S_z|-\rangle = +\frac{\hbar}{2}\Omega_0|-\rangle \end{cases} \quad [1.27a]$$

The eigenvalues of the Zeeman Hamiltonian W_Z are then written as:

$$\begin{cases} W_Z|+\rangle = W_+|+\rangle = -\frac{\hbar}{2}\Omega_0|+\rangle \\ W_Z|-\rangle = W_-|-\rangle = +\frac{\hbar}{2}\Omega_0|-\rangle \end{cases} \Rightarrow \begin{cases} W_+ = -\frac{\hbar}{2}\Omega_0 \\ W_- = +\frac{\hbar}{2}\Omega_0 \end{cases} \quad [1.27b]$$

The potential energies [1.16] are found since $\Omega_0 = \gamma H_0$.

Relations [1.20] show that in the magnetic field, the spin proton $I = 1/2$ has two energy levels. The energy difference between the upper level W_- (Ω_0 is positive) and the lower level W_+ is equal to:

$$\Delta E = W_- - W_+ = \hbar\Omega_0 \quad [1.28]$$

Let us move on to the demonstration of *Larmor precession*. To do this, let us denote by $\vec{v}$ the unit vector carried by the direction of the intrinsic magnetic moment $\vec{\mu}$. Under the action of the magnetic field $\vec{H}_0$, 1 the magnetic moment $\vec{\mu}$ is oriented in a direction making an angle θ with the vertical direction of the static field $\vec{H}_0$ (Figure 1.12).

In the magnetic field, the spin magnetic moment along $\vec{v}$ is:

$$\vec{\mu} = \mu_v \vec{v} \quad [1.29]$$

The direction of this vector is shown in Figure 1.12.

At $t = 0$, the spin eigenstate associated with the eigenvalue is $+\hbar/2$ [COH 77]:

$$|\Phi(0)\rangle = \cos\frac{\theta}{2}e^{i\varphi/2}|+\rangle + \sin\frac{\theta}{2}e^{-i\varphi/2}|-\rangle \quad [1.30]$$

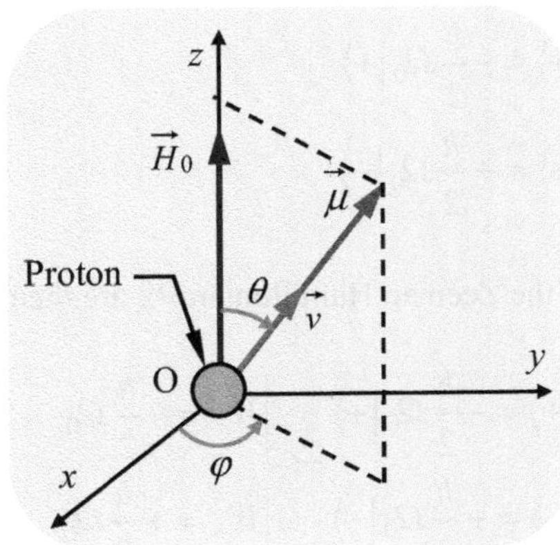

Figure 1.12. *Orientation of the spin $\vec{\mu}$ magnetic moment of the proton with respect to the direction of the magnetic field $\vec{H}_0$. For a color version of this figure, see www.iste.co.uk/sakho/nuclearphysics3.zip*

At time t, ket $|\Phi(t)\rangle$ is defined by:

$$|\Phi(t)\rangle = U(t)|\Phi(0)\rangle \quad [1.31]$$

In this relationship, $U(t)$ is the *devolution operator* given by:

$$U(t) = e^{-iHt/\hbar} \quad [1.32]$$

Replacing H by the Zeeman Hamiltonian W_Z gives:

$$U(t) = e^{-iW_Z t/\hbar} \quad [1.33]$$

Using [1.30] and [1.33], equation [1.31] is written as:

$$|\Phi(t)\rangle = U(t)|\Phi(0)\rangle = \cos\frac{\theta}{2}e^{i\varphi/2}\,e^{iW_Z t/\hbar}|+\rangle + \sin\frac{\theta}{2}e^{-i\varphi/2}\,e^{-iW_Z t/\hbar}|-\rangle \quad [1.34]$$

That is:

$$|\Phi(t)\rangle = \cos\frac{\theta}{2}e^{i\varphi/2}\,e^{-iW_+ t/\hbar}|+\rangle + \sin\frac{\theta}{2}e^{-i\varphi/2}\,e^{-iW_- t/\hbar}|-\rangle \quad [1.35]$$

Using the last equalities [1.27], the *state vector* [1.35] writing the evolution of spin 1/2 in the magnetic field $\vec{H}_0$ is written as:

$$|\Phi(t)\rangle = \cos\frac{\theta}{2} e^{i\frac{(\Omega_0 t+\varphi)}{2}} |+\rangle + \sin\frac{\theta}{2} e^{-i\frac{(\Omega_0 t+\varphi)}{2}} |-\rangle \qquad [1.36]$$

Let us consider

$$\varphi(t) = \Omega_0 t + \varphi \qquad [1.37]$$

Using [1.37], the state vector [1.36] takes the form:

$$|\Phi(t)\rangle = \cos\frac{\theta}{2} e^{i\frac{\varphi(t)}{2}} |+\rangle + \sin\frac{\theta}{2} e^{-i\frac{\varphi(t)}{2}} |-\rangle \qquad [1.38]$$

Since $\Omega_0 = \gamma H_0$, then:

$$\varphi(t) = \gamma H_0 t + \varphi \qquad [1.39]$$

Relation [1.39] shows that the magnetic field $\vec{H}_0$ introduces a phase shift between the coefficients assigned to the $|+\rangle$ and $|-\rangle$ eigenstates of the S_z observable. Therefore, in the presence of the magnetic field, the coordinates of the unit vector $\vec{v}(t)$ carried by the direction of the intrinsic magnetic moment $\vec{\mu}(t)$ are:

$$\vec{v}(t) = \begin{cases} \theta(t) = \theta \\ \varphi(t) = \Omega_0 t + \varphi \end{cases} \qquad [1.40]$$

The angle $\theta(t)$ between the directions of the unit vector $\vec{v}(t)$ and the magnetic field remains constant, while $\vec{v}(t)$ and therefore $\vec{\mu}(t) = \mu_v \vec{v}(t)$ rotate around the direction Oz of the field (Figure 1.13) with angular velocity:

$$\omega = \frac{d\varphi(t)}{dt} = \Omega_0 \qquad [1.41]$$

This magnetic field-induced rotational movement is called Larmor precession (Figure 1.13). The angular velocity Ω_0 is called the Larmor frequency (note that this frequency has the dimension of an angular velocity and not that of a frequency, as indicated by the formula [1.41]).

As shown in Figure 1.13, in the presence of an external magnetic field, each proton describes a double cone around the direction of $\vec{H}_0$, whose vertices meet at its center of gravity O.

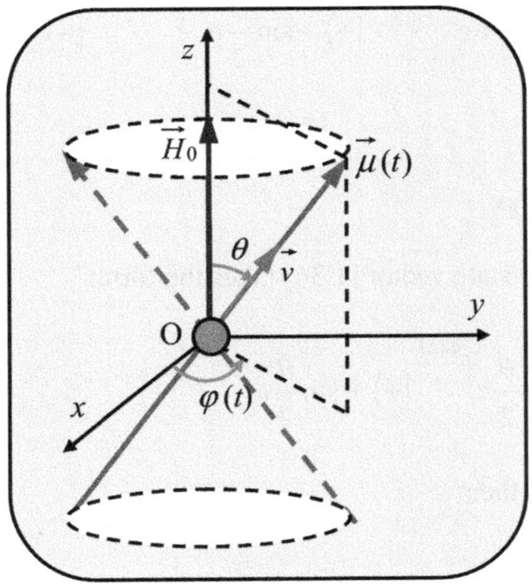

Figure 1.13. *Larmor precession of the spin magnetic moment of the proton subjected to an external magnetic field $\vec{H}_0$. The magnetic moment executes a double-cone motion around $\vec{H}_0$. For a color version of this figure, see www.iste.co.uk/sakho/nuclearphysics3.zip*

The protons in matter therefore rotate around $\vec{H}_0$ with a precession angle along the Oz axis and orient themselves in two possible directions. N_+ protons with magnetic moments parallel to the $\vec{H}_0$ field occupy the low-energy W_+ level, while N_- protons with magnetic moments antiparallel to $\vec{H}_0$ occupy the high-energy W_- level (Figure 1.11), with N_+ slightly greater than N_-. The difference in population $\Delta N = (N_+ - N_-)$ is linked to the magnetic force acting on each proton and to temperature. Although small (see Application 1.1), the difference ΔN is sufficient to produce a tissue-scale NMR signal and will induce the generation of a *macroscopic magnetization vector* $\vec{M}$ (M_{xy}, M_z) corresponding to the set of magnetic moments of the individual spins.

Since the protons' movements are out of phase, in equilibrium the macroscopic magnetization vector $\vec{M}$ remains stationary and aligned with the static field $\vec{H}_0$. Consequently, only the longitudinal component M_{z0}, known as *longitudinal magnetization*, remains, while the transverse component M_{xy0}, corresponding to transverse magnetization, is zero. The M_{z0} component increases with the proton concentration per unit volume in a voxel, called *proton density*, and with magnetic force. The proton density in a voxel is approximately equal to 2×10^{15} protons/mm^3 [EUS 15, REN 16].

1.3.5. *Excitation and magnetic resonance phenomenon*

At the steady state, the value of the macroscopic magnetization $\vec{M}$ oriented along Oz is not directly measurable, since it is very small compared to the field strength $\vec{H}_0$. To measure the value of the magnetization $\vec{M}$, it has to be tilted in the xOy plane (away from its equilibrium position) to give it a transverse component $\vec{M}_{xy}$. It is only through this component that the study of the macroscopic magnetization vector is possible, as it will no longer be "drowned" in the $\vec{H}_0$ field. This changeover, which corresponds to an excitation phase, takes place when external energy is supplied, using a second $\vec{H}_1$ *rotating magnetic field* or *excitation field*. This excitation is provided by a *pulsed radiofrequency* (RF) electromagnetic wave of frequency Ω_1.

When excited by the $\vec{H}_1$ field along the Ox axis, the magnetization vector $\vec{M}$ performs a double precession movement. The first movement takes place around the static field $\vec{H}_0$ at the angular frequency $\Omega_0 = \gamma_1 H_0$. The second movement takes place around the radiofrequency field at the angular frequency $\Omega_1 = \gamma_1 H_1$. During this double precession movement, the longitudinal component M_z decreases while the transverse component M_{xy} increases (Figure 1.14).

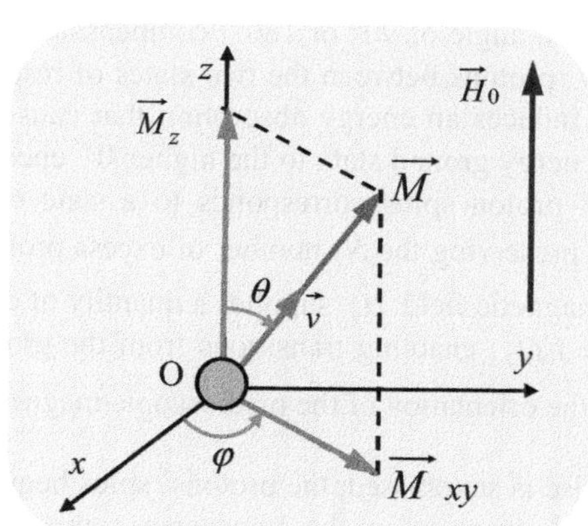

Figure 1.14. *Longitudinal M_z and transverse M_{xy} components of the macroscopic magnetization vector $\vec{M}$. For a color version of this figure, see www.iste.co.uk/sakho/nuclearphysics3.zip*

In practice, the RF wave is applied for short periods of a few milliseconds. This is referred to as an *RF pulse* or excitation, the value of the "flip" angle θ being a function of the duration of the RF pulse and the intensity of the $\vec{H}_1$ excitation field. The longer the RF wave is applied, the greater the angle θ. Most often, 90° or 180° pulses are used. A pulse of angle $\theta = 90°$ causes the magnetization vector $\vec{M}$ to tilt in the xOy plane (Figure 1.15a), while a pulse of angle $\theta = 180°$ (known as a 180° inversion pulse) causes $\vec{M}$ to invert (Figure 1.15b).

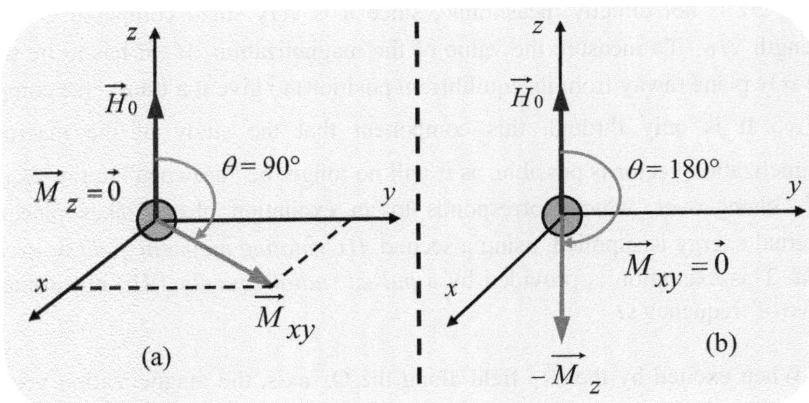

Figure 1.15. *Tilting the macroscopic magnetization vector $\vec{M}$ in the xOy plane (90° RF excitation) or on the Oz axis (180° RF excitation). For a color version of this figure, see www.iste.co.uk/sakho/nuclearphysics3.zip*

RF excitation (at an angle of 90° or 180°) compensates for the difference in the number of N_+ and N_- protons between the two states of respective energies W_+ and W_-. This excitation induces an energy absorption that causes the protons' spins to invert from the W_+ energy ground state to the higher W_- energy state. Therefore, the equilibrium state of proton spins corresponds to a state of dynamic equilibrium inducing transitions preserving the N_+ number of excess protons parallel to the $\vec{H}_0$ field. The rotating magnetic field $\vec{H}_1$ supplies a quantity of energy ΔE exactly equal to $\Delta E = W_- - W_+ = \hbar\Omega_0$, enabling transitions from the ground level to the higher level. This changes the orientation of the macroscopic magnetization $\vec{M}$.

When the RF pulse is suppressed, the protons' spins begin to move out of phase. This induces a rapid decrease in the transverse component M_{xy} and a gradual increase in the longitudinal component M_z of the magnetization vector $\vec{M}$. This phase of return to its equilibrium state, where it will execute a precession movement around the static field $\vec{H}_0$, is characterized by the simultaneous appearance of two relaxation phenomena: *longitudinal relaxation* lasting T_1 and *transverse relaxation* lasting T_2.

1.3.6. *Quantum model of the magnetic resonance phenomenon*

As we saw above, the spin magnetic moment of a proton immersed in a static magnetic field $\vec{H}_0$ takes on two possible orientations:

– The N_+ spins parallel to $\vec{H}_0$ occupy the ground energy level:

$$W_+ = -\frac{\hbar}{2}\Omega_0 = -\frac{\hbar}{2}\gamma_I H_0$$

– The N_- spins antiparallel to $\vec{H}_0$ occupy the higher energy level:

$$W_- = +\frac{\hbar}{2}\Omega_0 = +\frac{\hbar}{2}\gamma_I H_0$$

Moreover, we know that the rotating magnetic field H_1 supplies a quantity of energy exactly equal to:

$$\Delta E = W_- - W_+ = \hbar\Omega_0$$

As explained above, the energy ΔE is used to induce transitions from the ground level to the higher level. Knowing the energy of the radiofrequency electromagnetic wave $E = \Delta E = \hbar\Omega_1$, the phenomenon of magnetic resonance occurs when the frequency $\Omega_0 = \Omega_1$ [KAS 11, EUS 15, REN 16, IMA 16].

To illustrate the magnetic resonance phenomenon, let us assume that the proton difference $\Delta N = (N_+ - N_-) = 6$. In the presence of the static field $\vec{H}_0$ and in the absence of the exciting field $\vec{H}_1$, the six protons occupy the ground energy level W_+. At thermodynamic equilibrium, the N_+ population of the ground level is equal to 6 ($N_+ = 6$) and that of the upper level is empty ($N_- = 0$), as shown in Figure 1.16a. The application of an RF wave of increasing frequency causes transitions between the two levels. For a 90° pulse, half of the protons (three protons in this case) move from the ground to the upper level. The populations of the two energy levels W_+ and W_- are therefore equal (there are as many parallel protons as antiparallel protons at $\vec{H}_0$). This cancels out the longitudinal component M_z ($\vec{M}_z = \vec{0}$) of the magnetization vector $\vec{M}$, which tilts in the xOy plane (Figure 1.16b). The total spin magnetic moment is then zero, as the spins are out of phase.

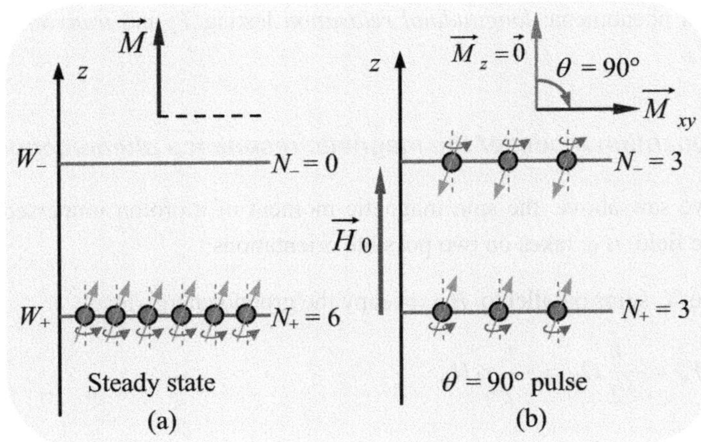

Figure 1.16. *Transition of half of the protons from the fundamental W_+ energy level to the higher W_- energy level following a 90° pulse of radiofrequency excitation. The longitudinal component of the magnetization vector is cancelled. M tilts in the xOy plane. The spins are globally out of phase, so that the total spin magnetic moment cancels out. For a color version of this figure, see www.iste.co.uk/sakho/nuclearphysics3.zip*

For a 180° pulse, the remaining half of the protons in the ground level move to the upper level. This reverses the longitudinal component of the magnetization vector $\vec{M}$: $\vec{M} = -\vec{M}_z$ (Figure 1.17).

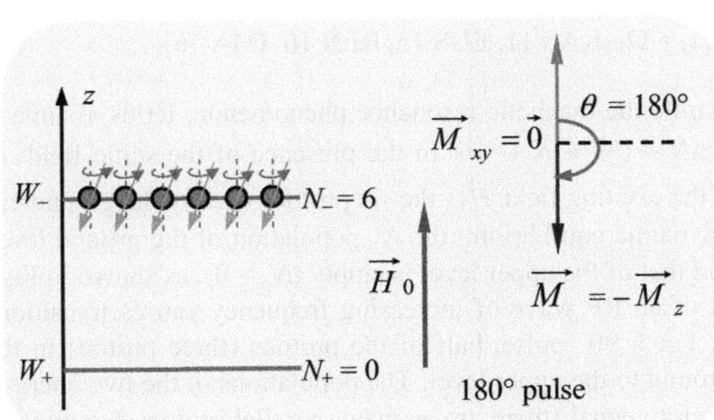

Figure 1.17. *Transition of all protons from the fundamental level to the higher energy level W_- following a 180° pulse radiofrequency excitation. There is a reversal of the longitudinal component of the magnetization vector M. For a color version of this figure, see www.iste.co.uk/sakho/nuclearphysics3.zip*

1.4. Magnetic resonance imaging (MRI)

1.4.1. *Nuclear magnetic resonance and medical imaging*

In 1971, it was established that the two *relaxation times* T_1 and T_2 of NMR were different between healthy and tumoral tissue, opening the way for numerous cancer diagnosis studies. In 1973, the American chemist Paul Lauterbur (1929–2007) proposed a medical imaging technique he called *Zeugmatography*. This technique is based on the spatial marking of spins using several magnetic field gradients, as well as on CT image reconstruction methods, enabling the acquisition of the first two-dimensional magnetic resonance images.

As explained above, NMR has practical applications in organic analysis and medicine, particularly in MRI. The latter offers a wide range of imaging options, including T_1-weighted sequences with or without contrast injection, T_2-weighted sequences, etc. MRI outperforms CT for most brain, ENT (*ear, nose and throat*) and pelvic tumors, thanks to better contrast and excellent spatial resolution. Nevertheless, dose calculations require the use of CT images, as well as registration between scanned and MR images [THI 07].

Today, MRI makes it possible to obtain 3D images in vivo, non-invasively, without irradiation, and with very high spatial resolution. A wide range of imaging modalities are possible with a single imaging device, including detailed anatomical images (structural MRI), functional activity images (functional MRI), diffusion images of water molecules (diffusion MRI), blood flow measurements (perfusion MRI), imaging of different metabolites (MRI spectroscopy), blood vessel imaging (MRI angiography), etc. During an MRI examination, several modalities can be used, enabling the multimodal study of different diseases.

1.4.2. *Magnetic resonance imaging techniques*

There are two types of MRI techniques (machine or imager): closed-field MRI (Figure 1.18a) and open-field MRI (Figure 1.18b).

The "closed" MRI technique is the most widespread and best-known configuration today. It consists, for the oldest MRI scanners, of a tunnel 60 cm in diameter and 2 m long, and for the most recent, 1.60 meters long. "Open" MRI emerged after closed MRI. Initially not very widespread, as open MRI technology improved, it was found to have advantages in human medicine, notably for people who could not benefit from this type of imaging in an enclosed space for practical reasons or to avoid general anesthesia. The magnet tunnel only concerns closed imagers and is the tunnel into which the patient is introduced. It has comfort functions (such as lighting and ventilation), as well as means of communication between the nursing staff and the patient (microphone and loudspeakers). Its diameter slightly varies according to manufacturer and model, but is approximately 60 cm (see: https://en.wikipedia.org/wiki/Magnetic_resonance_imaging, 2020).

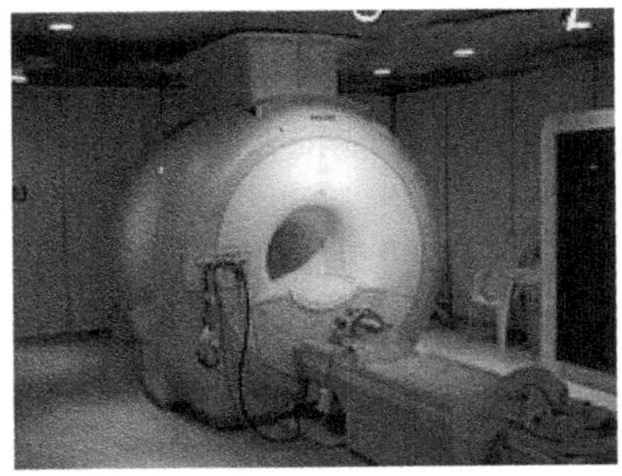

Figure 1.18a. *Closed-field MRI. For a color version of this figure, see www.iste.co.uk/sakho/nuclearphysics3.zip*

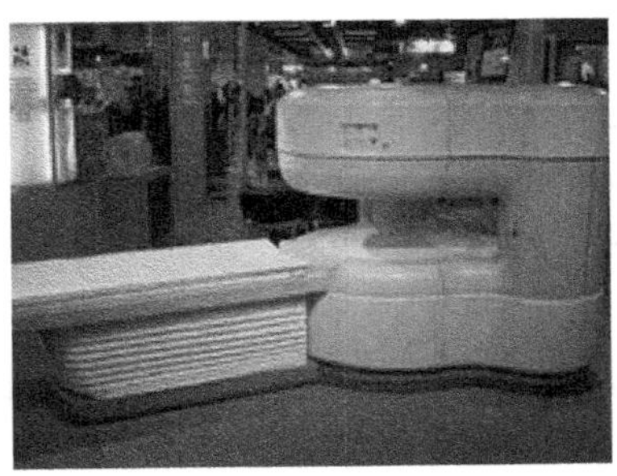

Figure 1.18b. *Open-field MRI (source: https://en.wikipedia.org/wiki/Magnetic_resonance_imaging, 2020). For a color version of this figure, see www.iste.co.uk/sakho/nuclearphysics3.zip*

MRI operates on four components: a superconducting magnet, cooled helium, radio waves and 3D *magnetic field gradients*. These components are described as follows (see: http://www.cinq.ulaval.ca/IRM).

Magnet

The main magnet creates the static magnetic field that gives rise to a measurable macroscopic magnetization. The most commonly used magnets are superconducting electromagnets. They consist of a coil rendered superconducting by liquid helium cooling, surrounded by liquid nitrogen. They deliver intense, homogeneous magnetic fields, but are expensive and require regular maintenance (notably topping up the helium tank). An MRI scanner operates continuously, so its magnetic field is always present, even when the scanner is not acquiring images. An essential characteristic of a superconducting magnet is the intensity of the field produced, measured in Tesla (T) or gauss. The magnetic field emitted by an MRI is extremely powerful. By comparison, the Earth's magnetic field is 5×10^{-5} T; in routine clinical practice, magnetic field strengths vary between 0.2 and 3.0 T. In research, magnets with intensities of 3.0 T to 7.0 T are used. For small animal imaging, MRI intensities can go well beyond 7.0 T, up to 21 T. The higher the intensity of an MRI magnet, the better the image resolution.

NOTE.–

Clinical medicine (etymologically from the Latin *clinice*, "medicine practiced near the patient's bed") is the study and treatment of diseases directly at the patient's bedside. It aims to establish a diagnosis through clinical examination, i.e. questioning and observation.

Cooled helium

Helium is used to cool the MRI magnet, giving it its superconductivity. At room temperature, helium is a gas; however, at very low temperatures, it becomes a liquid. The cooled helium used in an MRI scanner has a temperature close to absolute zero (between 2.5 and 3.5 Kelvin, i.e. around − 270 degrees Celsius). At this very low temperature, the magnet no longer has any resistance, and the electric current flows in a closed circuit, continuously through the magnet. This is the magnet's superconductivity. Helium must maintain its very low temperature at all times in order to remain liquid. This is why the MRI unit is permanently connected to the power supply, and why its superconducting magnet operates at all times.

Radiofrequency wave

In the MRI scanner, antennae emit radio waves whose energy is captured by the brain tissue, in particular by the hydrogen atoms present in this tissue. A radio wave is an electromagnetic wave with a frequency of less than 300 GHz, which corresponds to a wavelength in a vacuum greater than 1 millimeter. In addition to transmitters, the antenna also includes receivers to pick up the signal emitted by the tissues as they release energy. When radio waves are emitted into the MRI, the axis of rotation of the protons in the hydrogen atoms changes, and is no longer aligned with the magnetic field of the superconducting magnet. These protons, having absorbed the energy of the radio waves, are said to be in an excited state. When the radio waves stop transmitting, the protons relax and release the absorbed energy, returning to their initial position and gradually shifting phase. This induces a measurable radiofrequency signal which is picked up by the MRI. The time required for this realignment to the magnetic field depends on the type of tissue in which the hydrogen atom is located. This is what makes it possible to distinguish different tissues, notably white matter and gray matter, on MRI images.

Magnetic field gradients

The MRI unit contains gradient coils, positioned on three axes of a 3D plane. The electric current flowing through these coils produces a local distortion of the main magnetic field. This distortion is used for spatial image encoding. On each axis of a 3D plane (x, y and z), it enables a particular point to be located. Three gradient coils are used, one for each of the three planes of the 3D MRI images.

In what follows, we briefly explain the theoretical aspect of the NMR phenomenon. We then go on to study in detail the principle of its application to the study of biological tissues such as the human brain.

1.4.3. T_1 relaxation time and spin–lattice relaxation

When the radiofrequency wave pulse stops, the spins return to the ground state via microscopic transitions that depopulate the upper level. Two relaxation processes are observed, characterized by relaxation times T_1 and T_2 [EUS 15, REN 16, IMA 16].

In magnetic resonance imaging, the RF pulse specifically used corresponds to the Larmor frequency Ω of the hydrogen nucleus. As the protons (or spins) return from the higher energy level W_- to the fundamental energy level W_+, the energy absorbed during excitation is returned in the form of heat to the surrounding molecular medium, known as the *lattice*. For this reason, the relaxation process is also called *spin–lattice relaxation*. In addition, the energy released during relaxation gives an estimate of the number of protons or the amount of water present in the tissue under study.

To relate the behavior of the macroscopic magnetization vector to the relaxation times T_1 and T_2, we consider *magnetization regrowth* after the application of a 90° pulse to cancel the longitudinal component $\overrightarrow{M}_z$ of the tissue magnetization vector. During relaxation, there is a rapid phase shift of the spins. This implies a rapid decrease in the transverse component $\overrightarrow{M}_{xy}$ (Figure 1.19).

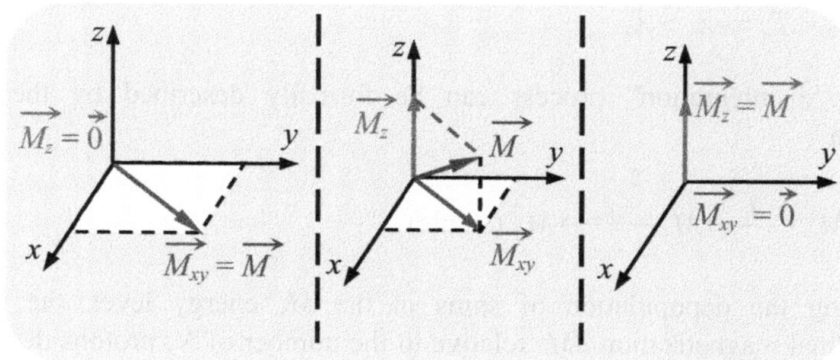

Figure 1.19. *Regrowth of longitudinal magnetization $\vec{M}_Z$ at the end of a 90° pulse. For a color version of this figure, see www.iste.co.uk/sakho/nuclearphysics3.zip*

In addition, the regrowth of longitudinal magnetization is characterized by the duration T_1, indicating the speed with which relaxation occurs.

In the following, we propose a simple methodology for a straightforward physical interpretation of the T_1 relaxation time, proceeding by analogy with the simple two-body spinning process (see Volume 1, section 4.2.2). Let M_0 be the maximum value of the longitudinal magnetization M_z at equilibrium after a 90° pulse has stopped.

Let M_+ be the longitudinal magnetization due to the X_+ nuclei (protons or spins + 1/2) present on the ground energy level W_+, and let M_- be the longitudinal magnetization due to the X_- nuclei (protons or spins − 1/2) present on the higher energy level W_-. The return of protons (or spins) from the upper energy level W_- to the ground energy level W_+ can be likened to a process of spin "decay" (or *depopulation* of the upper-level spins). The depopulation of an energy level corresponds to a decrease in longitudinal magnetization, while the settlement or accumulation of spins in an energy level corresponds to an increase in longitudinal magnetization.

Let λ_- be the *probability per unit* time that the longitudinal magnetization M_- "decays" or decreases. Similarly, let λ_+ be the probability per unit time that the longitudinal magnetization M_+ "decays" or decreases. T_- and T_+ are the respective "decay" periods of the M_- and M_+ magnetizations. It is clear that $\lambda_+ > \lambda_-$ since the W_- energy level empties much faster ($T_- > T_+$) than the W_+ energy level. Indeed, the depopulation times are:

$$T_+ = \frac{\ln 2}{\lambda_+}; T_- = \frac{\ln 2}{\lambda_-} \qquad [1.42a]$$

$$T_+ < T_- \Rightarrow \lambda_+ > \lambda_- \qquad [1.42b]$$

The "disintegration" process can be formally described by the following "filiation":

$$M_- \xrightarrow{\lambda_-} M_+ \xrightarrow{\lambda_+} M_\pm \text{ (stable)} \qquad [1.42c]$$

During the depopulation of spins in the M_+ energy level, the elementary longitudinal magnetization dM_+ relative to the number of N_+ protons decreases, and also increases by the amount $-\,dM_-$ resulting from the depopulation of spins in the M_- energy level. Let:

$$dM_+ = -\,\lambda_+ M_+ dt - dM_- \qquad [1.43]$$

For the M_- energy level, longitudinal magnetization decreases according to the differential equation:

$$dM_- = -\,\lambda_- M_- dt \qquad [1.44]$$

Using [1.44], equation [1.43] will be written as:

$$dW = -\,\lambda_+ M_+ dt + \lambda_- M_- dt \qquad [1.45]$$

Let W_{0-} be the value of the longitudinal magnetization relative to the number N_- protons. The *law of decay of the longitudinal magnetization M_- (t)* is written as (analogously to Rutherford and Soddy's empirical law equation [1.22], Volume 1 [SAK 22]):

$$M_-(t) = M_{0-} e^{-\lambda_- t} \qquad [1.46]$$

Carrying the solution [1.46] into the differential equation [1.45], we obtain:

$$\frac{dM_+(t)}{dt} + \lambda_+ M_+(t) = \lambda_- M_{0-} e^{-\lambda_- t} \qquad [1.47]$$

The solution to equation [1.47] is of the form:

$$M_+(t) = M_1 e^{-\lambda_- t} + M_2 e^{-\lambda_+ t} \qquad [1.48]$$

In equation [1.41], M_1 and M_2 are constants to be determined.

The solution of equation [1.48] is analogous to that of equation [4.48] in Volume 1 [SAK 22]. The *accumulation law for the number of spins* + 1/2 is then written as:

$$M_+(t) = \frac{\lambda_-}{\lambda_+ - \lambda_-} M_{0-}\left(e^{-\lambda_- t} - e^{-\lambda_+ t}\right) \quad [1.49]$$

We know that in the equilibrium state, the upper energy level W_- is empty, since all protons occupy the stable ground level. There is therefore no further depopulation. Therefore, $\lambda_- \rightarrow 0$. In addition $\lambda_+ >> \lambda_-$ (the ground state is stable), we obtain the following using [1.49]:

$$M_+(t) \approx \frac{\lambda_-}{\lambda_+} M_{0-}\left(1 - e^{-\lambda_+ t}\right)$$

Let us say that:

$$(\lambda_-/\lambda_+)\, M_{0-} = M_0;\ T_1 = 1/\lambda_+ \quad [1.50]$$

The final result is obtained by assuming $M_+ = M_z$:

$$M_z(t) = M_0\left(1 - e^{-t/T_1}\right) \quad [1.51]$$

Using [1.44], the physical meaning of T_1 relaxation is easily clarified.

– For $t = T_1$, $W_z\,(T_1) = W_0(1 - e^{-1}) = 0.63\ W_0 = 63\%\ W_0$ [1.52a]

– For $t = 5T_1$, $W_z\,(5T_1) = W_0(1 - e^{-5}) = 0.99\ W_0 = 99\%\ W_0$ [1.52b]

The results [1.52] recall the phenomenon of charging a capacitor of capacitance C through an ohmic conductor of resistance R (see Volume 1 in [SAK 22], section 4.2.2, formula [4.90]). The relaxation T_1 is analogous to the time constant $\tau = RC$ of the circuit (R, C).

Therefore, by definition, relaxation T_1 is the characteristic *time constant* for a given tissue. It corresponds to 63% of the regrowth of the longitudinal magnetization M_z, i.e. the time taken for M_z to reach 63% of its equilibrium value M_0 after a 90° pulse has stopped. The shorter the time constant T_1, the faster the longitudinal relaxation. At $t = 5\ T_1$, longitudinal magnetization has practically reached its maximum value [2.52]. The order of magnitude of spin–lattice relaxation T_1 for biological tissue is 500–1,000 ms [REN 16].

Now let us study the time variation of longitudinal magnetization. The curve obtained is shown in Figure 1.20.

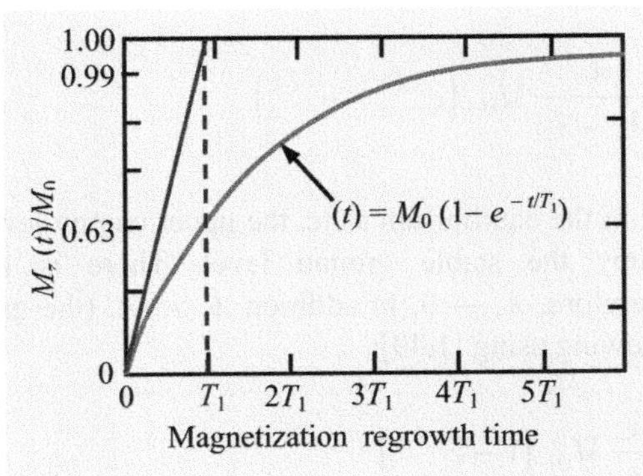

Figure 1.20. *Time variation of longitudinal magnetization M_z after a 90° pulse has stopped. The relaxation T_1 characterizing the time constant of the tissue studied corresponds to 63% of the regrowth of the longitudinal magnetization M_z. For a color version of this figure, see www.iste.co.uk/sakho/nuclearphysics3.zip*

In tissue environments, molecules are constantly subjected to various motions (translation, rotation, collision) characterized by the *frequency of molecular collisions* noted ν_c. The *condition of resonance* occurs when the frequency $\nu_0 = \Omega_0/2\pi$ of protons in the $\vec{H}_0$ field is close to the frequency of molecular collisions ν_c. In this case, the protons transfer their energy to the surrounding biological medium.

It should be noted that several parameters, such as the molecular structure of the tissue and the solid or liquid state of the material, influence the value of T_1 [REN 16].

For media containing large molecules such as lipids, for example, molecular movements are fairly slow ($\nu_c \approx \nu_0$). Consequently, energy exchanges have maximum efficiency and T_1 is short. We deduce that T_1 relaxation of fat is short. For crystalline media, molecular movements are extremely slow, $\nu_c < \nu_0$. Consequently, T_1 relaxation is long.

In liquids, T_1 relaxation is longer than in solids. In the case of pure water containing small, fast-moving molecules (H_2O molecules), the frequency of molecular collisions is much higher than the Larmor frequency ($\nu_c > \nu_0$). The relaxation time T_1 is therefore long. Consequently, the increase in water content within a tissue due to, for example, a tumor (increase in tissue volume, without specifying the cause), edema (tissue swelling due to the presence of an abnormal quantity of fluid) or *infarction* (see myocardial infarction, Chapter 2, section 2.7.8) will lengthen the *tissue time* constant T_1. Conversely, the presence of large-molecule proteins in a fluid will shorten T_1 ($v_c \approx v_0$).

T_1 also decreases with medium viscosity, sample temperature and the presence of paramagnetic substances (contrast agents, see note after Figure 1.32).

1.4.4. *T_2 relaxation time and spin–spin relaxation*

As shown in Figure 1.19, as soon as the RF pulse stops, the transverse magnetization $\overrightarrow{M}_{xy}$ decreases as the longitudinal magnetization $\overrightarrow{M}_z$ grows back. This decay corresponds to *transverse relaxation,* characterized by a time constant T_2. It is assumed that the transverse magnetization decays faster than the longitudinal magnetization that regrows, so that the tissue has a time constant T_2 that is always less than or equal to the time constant T_1 characteristic of the regrowth of $\overrightarrow{M}_z$. This transverse relaxation resulting from the *spin–spin interaction* of protons is referred to as *spin–spin relaxation* obeying the exponential decay law:

$$W_{xy}(t) = W_0 e^{-t/T2} \qquad [1.53]$$

According to the law of decay [1.46], for $t = T_2$, we obtain:

$$W_{xy}(T_2) = W_0 e^{-1} = 0.3678\ W_0 \approx 0.37\ W_0 \qquad [1.54]$$

Therefore, the time constant T_2 corresponds to the time for which the value of the transverse magnetization is only 37% of its maximum value M_0 (Figure 1.20) after the 90° pulse has stopped.

Figure 1.22 shows the time evolution of the longitudinal and transverse components of the macroscopic magnetization after 90° RF excitation. In particular, T_1 is much larger than T_2.

In physiological tissues, and at the magnetic field strengths H_0 at which these tissues can be imaged, the relaxation time T_1 is generally much greater than T_2 (Figure 1.22). T_1 and T_2 relaxation times for healthy physiological tissue are given in the literature. Their values depend on the condition of the tissue (e.g. tissue examined in vivo does not give the same results as the same tissue examined in vitro), the temperature, the action of any contrast agents used and the static magnetic field in which the tissue is placed [SHP 74, POS 85, NEL 87].

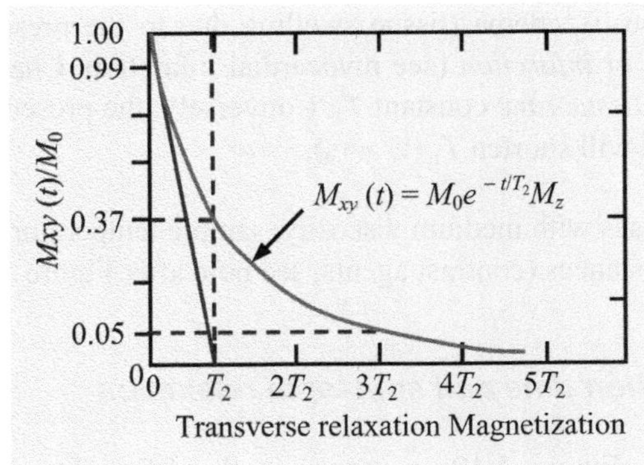

Figure 1.21. *Transverse relaxation corresponds to the decay of transverse magnetization M_{xy} after a 90° pulse has stopped. The T_2 time constant characteristic of the tissue studied corresponds to 37% of the M_{xy} during relaxation. For a color version of this figure, see www.iste.co.uk/sakho/nuclearphysics3.zip*

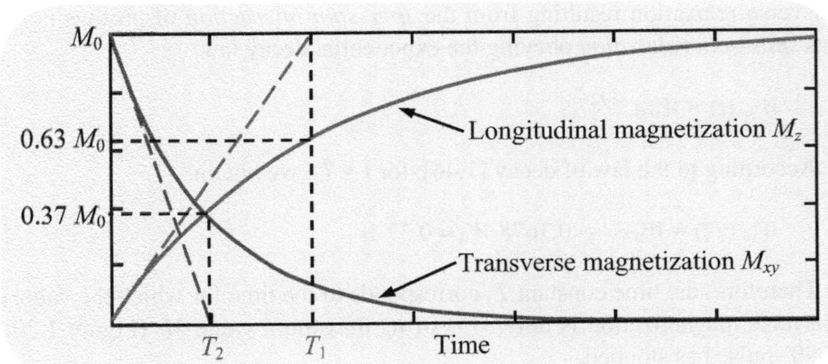

Figure 1.22. *Evolution with time of the longitudinal M_z and transverse M_{xy} components of the macroscopic magnetization vector after excitation by a 90° RF wave. For a color version of this figure, see www.iste.co.uk/sakho/nuclearphysics3.zip*

Broadly speaking, the effect of the magnetic field on relaxation times can be summarized by saying that T_1 increases rapidly with increasing H_0, while T_2 is relatively constant. As each tissue has its own microscopic properties depending on its composition, T_1 and T_2 are therefore characteristic of the microscopic structure of the tissue [ROQ 13]. Consequently, there is a real clinical interest in knowing the T_1 and/or T_2 values of a tissue in order to image it. Therefore, the different intensities present in an MRI image are derived from T_1 and T_2 values. Table 1.3 shows some T_1 and T_2 values for selected tissues [STA 05].

Tissue	T_1 **(ms)**	T_2 **(ms)**
Heart	2,200–2,400	500–1,400
Blood	920	100
Liver	240–250	60–80
Skeletal muscle	1,350	50
White matter	1,350	200
Gray matter	860–900	50

Table 1.3. *T_1 and T_2 relaxation time values for selected tissues [EUS 15]. Measurements were made ex vivo at $H_0 = 3$ T and $T = 37$°C. Note that $T_1 > T_2$*

NOTE.–

The in vitro concept (from the Latin "under glass") applies to any experimental activity carried out on microorganisms, organs or cells outside their natural context (outside the living organism or living cell) and under defined, controlled conditions (e.g. in vitro fertilization). The ex vivo concept (from the Latin "outside the living") defines biological tests carried out outside the organism. These are cell cultures made from cells extracted from an organism. *Gray matter* denotes the tissue of the central nervous system (CNS), the part of the nervous system comprising the brain and the spinal cord (old nomenclature). Under the microscope, gray matter appears darker than the rest of the nervous tissue, known as *white matter*.

1.4.5. *Relaxation time T_2* and free induction decay (FID)*

As explained above, during the relaxation process after the 90° pulse has stopped, the transverse component $\vec{M}_{xy}$ of the macroscopic magnetization vector $\vec{M}$ decreases faster than the regrowth of its longitudinal component $\vec{M}_z$. While precessing around $\vec{H}_0$, the end of the macroscopic magnetization vector spirals on a cone-shaped surface (Figure 1.23).

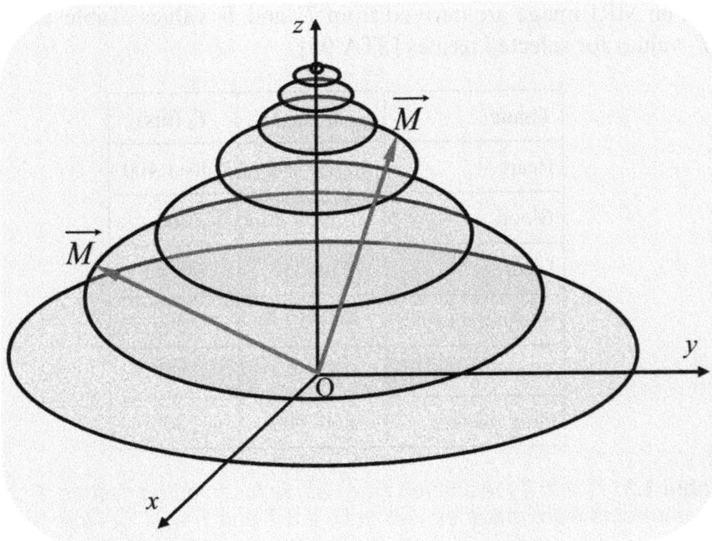

Figure 1.23. *Trajectory of the end of the macroscopic magnetization vector* $\vec{M}$ *after the 90° pulse has stopped. For a color version of this figure, see www.iste.co.uk/sakho/nuclearphysics3.zip*

In a voxel with a proton density of approximately 2×10^{15} protons/mm^3, all spins are immersed in the applied magnetic field $\vec{H}_0$. As we have seen in section 2.3.5, at the steady state, the value of the macroscopic magnetization oriented along Oz is not directly measurable since it is very small compared to the field strength $\vec{H}_0$. Only the transverse component $\vec{M}_{xy}$ is measurable.

To measure the NMR signal emitted by selected spins during the relaxation phase of the transverse component of magnetization, the phenomenon of *electromagnetic induction* is exploited. By rotating a magnet in front of a coil, an induced current appears at the coil terminals, as shown in Figure 1.24 [SAK 18]. A straight magnet moved along the axis of a coil (Figure 1.24a) or away from it (Figure 1.24b) generates an electric current of intensity i, the direction of which depends on the direction of movement of the magnet. The phenomenon that gives rise to the induced current is called electromagnetic induction.

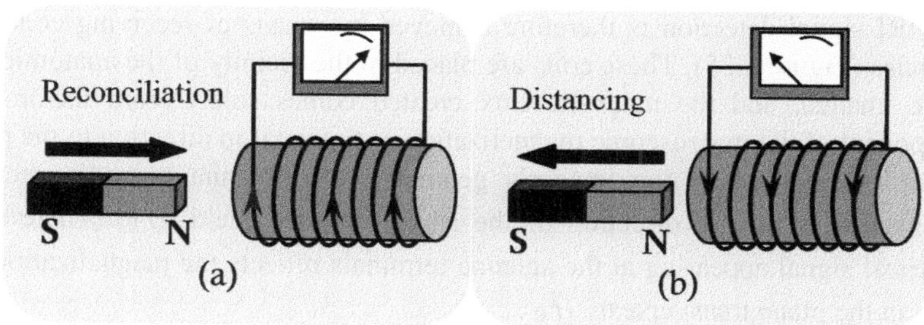

Figure 1.24. *(a) Magnet drawn closer together along the horizontal axis of a coil: an electric current is generated. (b) Magnet separated along the horizontal axis of a coil: an electric current is generated. For a color version of this figure, see www.iste.co.uk/sakho/nuclearphysics3.zip*

The change in magnetic field flux Φ across the coil gives rise to an electric voltage known as the *induced electromotive force (EMF)*, denoted by *e*. By virtue of *Faraday's law*, the EMF is related to the variation in magnetic field flux, according to the expression:

$$e = -\frac{d\Phi}{dt} \qquad [1.55]$$

Let us project the end of the macroscopic magnetization vector $\vec{M}$ onto the xOy plane. During the procession of $\vec{M}$ around the magnetic field $\vec{H}_0$, the end of the transverse component $\vec{M}_{xy}$ describes a spiral on the xOy while decreasing (Figure 1.25).

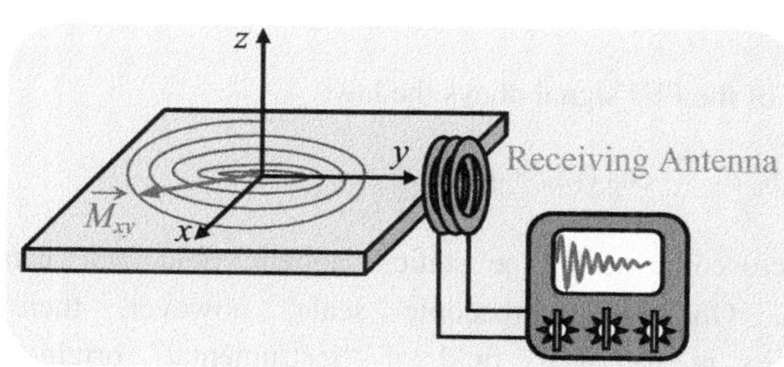

Figure 1.25. *Trajectory of the end of the macroscopic magnetization vector after the 90° pulse has stopped. For a color version of this figure, see www.iste.co.uk/sakho/nuclearphysics3.zip*

MRI signal detection is therefore achieved by means of receiving coils, called antennas (Figure 1.25). These coils are placed in the vicinity of the anatomical zone to be imaged, and the magnetic flux created comes solely from the transverse component of the macroscopic magnetization, orthogonal in direction to the plane of the coils. Because of the imager's geometry, the antenna plane is necessarily collinear with the Oz direction of the static magnetic field $\vec{H}_0$. Therefore, the electrical signal appearing at the antenna terminals reflects the magnetization of the spins in the plane transverse to $\vec{H}_0$.

As shown in Figure 1.25, rotation of the transverse component $\vec{M}_{xy}$ induces a magnetic field corresponding to a radiofrequency wave. This phenomenon corresponds to a signal known as *free induction decay* (FID). FID is demonstrated by collecting the signal via a coil or receiving antenna placed in the xOy plane. The signal is then transformed into a measurable electrical signal, giving a sinusoid damped by a time exponential in T_2 (Figure 1.26).

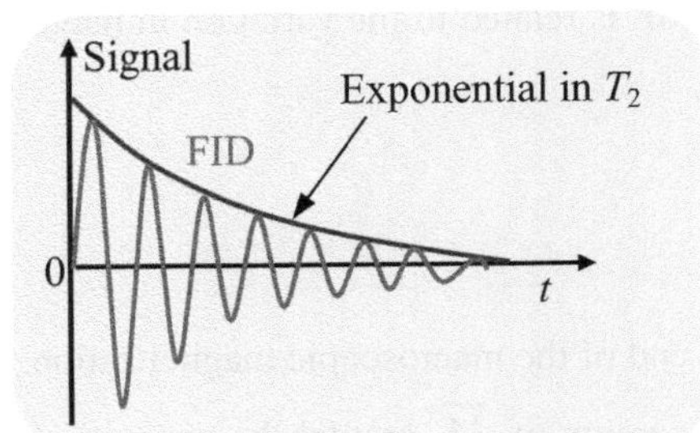

Figure 1.26. *Free induction decay (FID) signal transformed into a measurable electrical signal giving a sinusoid damped by a time exponential in T_2. For a color version of this figure, see www.iste.co.uk/sakho/nuclearphysics3.zip*

The decay of the FID signal obeys the law:

$$S(t) = S_0\, e^{-t/T2} \cos(\Omega_0 t + \varphi) \qquad [1.56]$$

On a macroscopic scale, the static magnetic field $\vec{H}_0$ can be considered homogeneous. On the microscopic scale, however, there are constant inhomogeneities in the $\vec{H}_0$ field of "instrumental" origin. These *inherent inhomogeneities* in the $\vec{H}_0$ field lead to an acceleration in spin dephasing.

When the inhomogeneities of $\vec{H}_0$ are taken into account, the decay of the FID signal is no longer a decreasing exponential in T_2. In fact, the observed FID signal is linked to the inhomogeneities of the $\vec{H}_0$ field of molecular origin (T_2), to which are added the inherent inhomogeneities of the $\vec{H}_0$ magnetic field of instrumental origin. To represent the conjunction of these two effects, we use the symbol T_2*. The observed FID signal then decays faster than expected, according to a T_2* exponential (Figure 1.27).

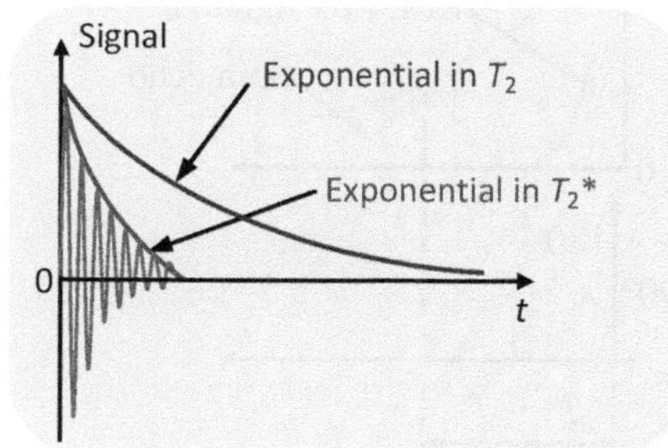

Figure 1.27. *FID signal decays exponentially in T_2* and T_2. T_2* decay is due to the inhomogeneities of the $\vec{H}_0$ field of molecular origin, plus the inherent inhomogeneities of the $\vec{H}_0$ field of instrumental origin. For a color version of this figure, see www.iste.co.uk/sakho/nuclearphysics3.zip*

1.4.6. *Spin echo sequence*

The two main medical imaging sequence methods that have been developed to cover a very wide range of MRI applications are the *spin echo sequence*, which is characterized by the presence of a 180° RF pulse, and the *gradient echo sequence* (see section 1.4.7).

As we know, at the end of the 90° pulse, the cumulative effects of the inhomogeneities of the $\vec{H}_0$ field of instrumental and molecular origin cause the FID signal to decay exponentially in T_2*, faster than expected. In order to access the true T_2 relaxation time by eliminating the $\vec{H}_0$ field's own inhomogeneities, we take advantage of the so-called *spin echo sequence*, characterized by its *echo time TE*, which defines the time elapsed between the 90° RF pulse stop and signal collection.

We then assume that an FID induction decay signal has been created at time $t = 0$, using a 90° pulse. Then, after a time T equal to *half the echo time* ($T = TE/2$), a 180° pulse is applied when the FID signal has practically disappeared. This causes an inversion of the longitudinal component of the magnetization vector, corresponding to an inversion of the spin phase shifts. This inversion occurs without any change in the direction or speed of spin precession. An induction signal, known as the *spin echo signal* or echo of the initial FID signal, appears once again (Figure 1.28).

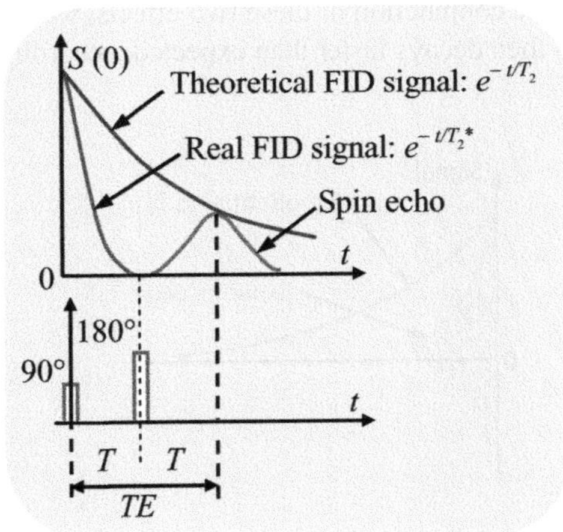

Figure 1.28. *Spin echo signal or echo of initial FID signal. For a color version of this figure, see www.iste.co.uk/sakho/nuclearphysics3.zip*

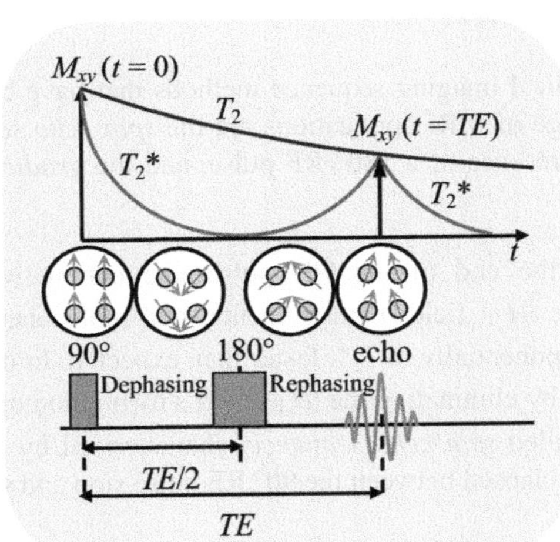

Figure 1.29. *Spin dephasing and rephasing during spin echo signal evolution. For a color version of this figure, see www.iste.co.uk/sakho/nuclearphysics3.zip*

Figure 1.29 shows the phase shift and rephasing of spins during the evolution of the spin echo signal.

The appearance of the spin echo signal can be described in three phases:

1) At the initial instant $t = 0$, the spins are in phase and oriented in the direction of the static field $\vec{H}_0$. At this point, a 90° pulse is applied, causing the tissue magnetization vector to shift in the xOy plane (cancellation of the longitudinal component $\vec{M}_z$: $\vec{M} = \vec{M}_{xy}$).

2) Due to the inhomogeneities of the $\vec{H}_0$ field, the rotation of $\vec{M}_{xy}$ induces an FID signal that decays more rapidly according to an exponential in T_2*. This decay is accompanied by a rapid dephasing of the proton spins.

3) When $\vec{M}_{xy}$ cancels at time $t = TE/2$, a 180° pulse is applied. The transverse component increases in the interval $TE - TE/2$. During this growth, there is a progressive rephasing of the spins, followed by the appearance of the spin echo signal collected at time $t = TE$. Beyond this point, the transverse magnetization decreases again according to an exponential in T_2*.

When the spins are in phase, transverse magnetization becomes maximum. A second 90° pulse can be applied to resume the cycle. The time interval between two 90° pulses, i.e. between two elementary cycles, is called the *repetition time* (*TR*).

The spin echo signal Sse obtained in a voxel is a function of M_0, T_1 and T_2 for tissue properties, and of *TE* and *TR* for acquisition parameters. The expression of the spin echo signal from each voxel of the selected slice is given by the equation:

$$Sse = M_0 \frac{\left(1 - 2e^{-(TR+TE)/T_1}\right) + e^{-TR/T_1}}{1 + \cos\theta\, e^{-TR/T_1}} \sin\theta\, e^{-TE/T_2} \qquad [1.57]$$

The spin echo sequence is carried out for a 90° pulse (θ = 90°), so equation [1.57] can be summarized as follows:

$$Sse = M_0 \left(1 - 2e^{-(TR+TE)/T_1} + e^{-TR/T_1}\right) e^{-TE/T_2} \qquad [1.58]$$

In practice, the *TR* repetition time corresponds to the passage from one line to the next (a complete "image" cycle corresponds to 128 or 256 lines) [REN 16]. For a long *TR*, so that *TR* >> *TE*, the spin echo sequence equation is approximately [DUC 15]:

$$Sse = M_0\left(1 - e^{-TR/T_1}\right)e^{-TE/T_2} \qquad [1.59]$$

The $TR >> TE$ condition is achieved in the case of a long T_2-weighted *proton density sequence* (see below). This is the case, for example, with *pathological phenomena* that generally lengthen T_2, giving a T_2 hypersignal. This type of long TR sequence (2,000 ms) also allows proton density weighting, which makes contrast a little more difficult to understand. In fact, the sequence generally comprises two asymmetrical echoes: a first early echo with a short TE (15–20 ms), weighted in proton density, and a second late echo with a long TE (100–120 ms) weighted in T_2 [REN 16]. For TR = 2,000 ms and TE = 20 ms or 120 ms, the approximation [1.59] is valid.

Equations [1.58] and [1.59] show that the spin echo signal depends on the transverse relaxation time T_2 (not T_2*) of the sample. This reflects the fact that the influence of $\vec{H}_0$ heterogeneities is eliminated at time $2T = TE$. However, these heterogeneities are not completely eliminated, as the 180° pulse only eliminates phase shifts due to $\vec{H}_0$ field inhomogeneities, which are constant, but does not eliminate random phase shifts due to field inhomogeneities of molecular origin, which are irreversible (spin–spin relaxation) [REN 16]. Figure 1.30 schematically illustrates the benefits of applying a 180° pulse to overcome the inherent inhomogeneities of the $\vec{H}_0$ static field (mirror image of inhomogeneities).

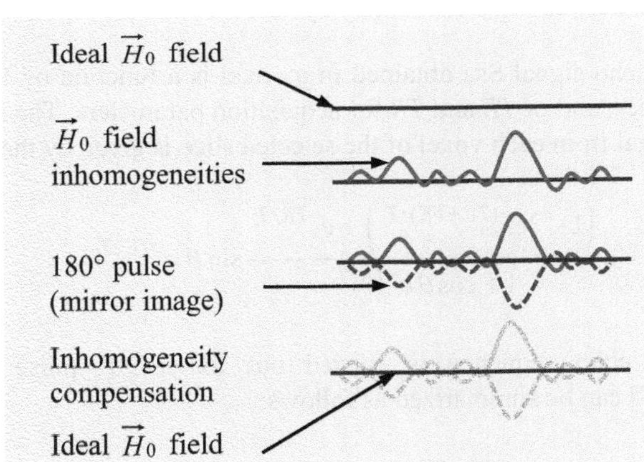

Figure 1.30. *Schematic illustration of the benefits of applying a 180° pulse to overcome the inherent inhomogeneities of the instrumental $\vec{H}_0$ field. The 180° pulse produces inverted inhomogeneities, giving an image figure in relation to a horizontally positioned mirror containing the direction of the $\vec{H}_0$ field. For a color version of this figure, see www.iste.co.uk/sakho/nuclearphysics3.zip*

NOTE.–

Static magnetic fields used in MRI are of the order of a few teslas, mainly 1.5 T and 3 T. Some MRIs dedicated to research use higher fields (e.g. $H_0 = 7$ T). These fields are very high: in comparison, the Earth's magnetic field is of the order of 50 μT. The RF field strength is 64 MHz for a magnetic field of 1.5 T and 128 MHz for a field strength of 3 T [INS 18].

In MRI, image contrast depends on intrinsic tissue parameters (T_1, T_2, proton density or proton concentration per unit volume), as well as sequence parameters such as echo time *TE* and repetition time *TR* to highlight tissue characteristics.

T_1 and T_2 weighting

In MRI, T_1-*weighted images* follow the growth of the longitudinal component of macroscopic magnetization, while T_2-*weighted images* follow the decay of the transverse component of macroscopic magnetization. For biological tissues, the order of magnitude of T_1 is 500–1,000 ms, while that of T_2 is 50–100 ms.

Tissues with a naturally short T_1 have a high fat or protein content. Fat is easily identifiable on MRI. For example, at T_1 = 150 ms and magnetic field intensity H_0 =1.5 T, fat appears as a *hypersignal*. This allows us, for example, to shed light on the fatty or nonfatty nature of a lesion in cases of diagnostic doubt. In this precise case, a fat signal suppression sequence usually solves the problem by showing a decrease in the lesion's signal.

Proteins present in biological fluids modify their T_1 and T_2 relaxation times. Liquids containing negligible amounts of protein have a signal close to that of water, black in T_1 and white in T_2. For the majority of tissues, water is mainly located in the intracellular environment. In some tissues, such as those with an important secretory function, water is very abundant in the extracellular medium or within microtubules. As the protein content of fluids increases, the T_2 signal initially increases slightly, then decreases. On the contrary, the T_1 signal increases progressively, peaking when the fluid contains 25% protein and then decreasing beyond that [REN 16]. Joint analysis of T_1- and T_2-weighted sequences can therefore be used to assess the protein content of a given lesion.

Proton density weighting or ρ

As explained at various points in this section, during the relaxation process after the 90° pulse has stopped, the transverse component $\overrightarrow{M}_{xy}$ of the macroscopic magnetization vector decreases faster than its longitudinal component $\overrightarrow{M}_z$, until it reaches its initial maximum value $\overrightarrow{M}_{z0}$ at equilibrium. The $\overrightarrow{M}_{z0}$ value increases with *proton density*. MRI contrast therefore expresses differences in proton density. However, this contrast is generally low, as proton density differences in biological tissues are not very high. T_2-weighted sequences are better contrasted than T_1-weighted sequences in ρ proton density. However, a weaker signal is obtained due to the fact that measurements are made late on the T_2 signal attenuation curve.

Influence of repetition time TR and echo time TE

As we know, the *TR* repetition time is the time interval between two 90° pulses. The *TR* determines the contrast or T_1 weighting of a given sequence [REN 16].

– The shorter the *TR* repetition time, the greater the T_1 contrast (Figure 1.31a). In this case, the *sequence is said to be* T_1*-weighted.* The tissue with the shortest T_1 then gives the highest signal.

– On the contrary, the *longer the TR* repetition time, the lower the T_1 contrast (Figure 1.31b), as the T_1 differences between the tissues are less perceptible. In this case, the *sequence is* T_1*-weighted.*

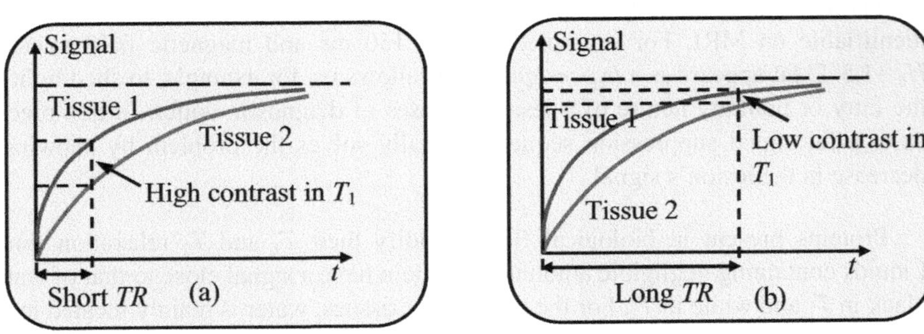

Figure 1.31. *Influence of TR repetition time: (a) T_1-weighted sequence; the shorter the TR, the higher the T_1 contrast. (b) T_1-weighted sequence; the longer the TR, the lower the T_1 contrast. For a color version of this figure, see www.iste.co.uk/sakho/nuclearphysics3.zip*

Remember that the echo time *TE* is the time elapsed between the 90° RF pulse being stopped and the signal being collected. Therefore, it determines the moment at which the signal is measured on the T_2 decay curve (time during which the signal is allowed to decay in T_2 before being measured). If the *TR* conditions the contrast or T_1 weighting of a given sequence, the contrast or T_2 weighting of a sequence is conditioned by the *TE*.

– The *shorter the TE* echo time, the lower the T_2 contrast. In this case, the sequence is T_2 *deweighted* (Figure 1.32a); the two tissues (1 and 2) cannot be distinguished by their T_2 relaxation time, as the differences in decay rate do not have time to be expressed.

– The *longer the TE* echo time, the stronger the T_2 contrast. *This is said to be T_2-weighted* (Figure 1.32b).

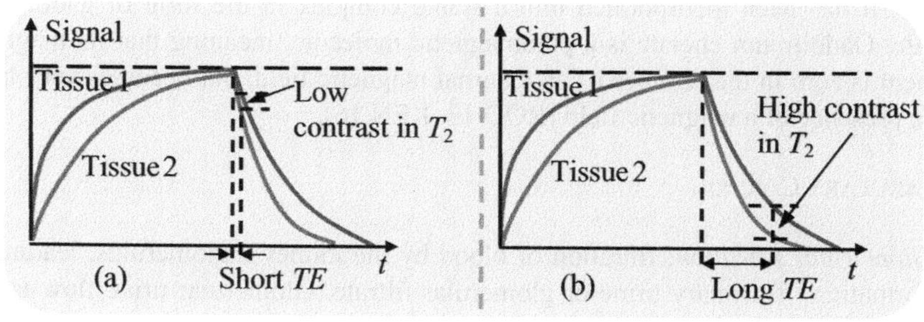

Figure 1.32. *Influence of TE echo time. (a) T_2-weighted sequence; the stronger the T_2 contrast, the longer the TE. (b) T_2-weighted sequence; the weaker the T_2 contrast, the shorter the TE. For a color version of this figure, see www.iste.co.uk/sakho/nuclearphysics3.zip*

NOTE.– Contrast agents

As previously mentioned, the contrast of MRI images depends on intrinsic tissue parameters (T_1, T_2, proton density ρ), as well as on sequence parameters *TR* (conditioning contrast or T_1 weighting) and *TE* (conditioning contrast or T_2 weighting).

Tissue characterization can be enhanced by the use of *contrast agents*. The action of MRI contrast agents is indirect: the agent itself is not visible, but its influence on the relaxation times of surrounding water protons can be detected. More precisely, MRI contrast agents shorten the T_1 and T_2 relaxation times of tissues, with one or other of these effects predominating depending on the agent used or its concentration. Contrast agents with a predominant T_1 effect, by decreasing the T_1 relaxation time of tissues, lead to an increase in the signal of the structure with which they are in contact: these are known as *"positive" contrast agents*. Contrast agents with a predominant T_2 effect reduce the signal of structures with which they come into contact: these are known as *"negative" contrast agents*. These products are administered intravenously at a usual dose of 0.2 mL/kg body weight. Their pharmacokinetics are similar to those of iodinated contrast agents (used in X-ray imaging). After injection, they are rapidly transferred from the vascular sector to the interstitial space. They are then excreted by the kidney via *glomerular filtration*. *Gadolinium* is a metal of the lanthanide or rare-earth class. With its seven single electrons, it has the highest relaxivity, and consequently the greatest effectiveness in modifying relaxation times. At low concentrations, its main effect is to shorten the T_1 relaxation time of tissues, leading to an increase in the signal of tissues with which it is in contact (positive contrast agent). However, as its concentration increases, the T_2 effect becomes predominant, resulting in a decrease in signal. Free gadolinium (gadolinium salt) is highly toxic. To mask its harmful effects, it has been incorporated into a stable complex in the form of gadolinium chelate. Gadolinium chelate is a paramagnetic molecule, meaning that its magnetic moment is zero in the absence of an external magnetic field, but becomes very large in the presence of a magnetic field [ROQ 13, REN 16].

VOCABULARY CORNER.–

Glomerular filtration: filtration of blood by the kidney's glomerulus, leading to the formation of primary urine or glomerular filtrate. Glomerular urine flow is 180 liters/24h. Overall kidney function is quantified by glomerular filtration [INS 20]. Each kidney comprises around 1 million glomeruli located in the renal cortex (tissue just below the renal capsule).

Glomerular filtration rate (GFR): corresponds to the volume of plasma filtered by the kidney per unit of time. It depends on the number of functional nephrons and the GFR per nephron. The average number of nephrons per kidney is 1 million, but there is considerable inter-individual variation (from 400,000 to 1,300,000). The number of nephrons is partly genetically determined and therefore appears to be roughly correlated with birth weight and parental height [COL 16].

The *nephron* is the structural and functional unit of the kidney. It enables the formation of urine. Each of the two human kidneys contains around 1,000,000 nephrons.

1.4.7. *Gradient echo sequence*

Unlike the spin echo sequence, the *gradient echo sequence* does not have a 180° pulse. The first radiofrequency pulse is generally less than 90°. The signal equation for the gradient echo sequence is:

$$Sse = M_0 \frac{\left(1 - e^{-TR/T_1}\right)}{1 - e^{-TR/T_1} \cos\theta} \sin\theta \, e^{-TE/T_2^*} \quad [1.60]$$

The tilting angle $\theta < 90°$ will reduce the portion of magnetization tilted. The result is a faster return to equilibrium. The repetition time becomes shorter. The absence of a 180° pulse makes the gradient echo sequence sensitive to inhomogeneities in the $\vec{H}_0$ magnetic field. Images obtained in this case are T_2*-weighted. This reduces the *TR*. The main advantage of the gradient echo sequence is that it enables faster image acquisition than the spin echo sequence.

1.4.8. *The slice plane and the slice selection gradient*

MRI signal measurement alone does not form an image. To achieve this, it is necessary to carry out a *coding* operation, associating a position to each of the signals to form the image. This coding is carried out using a succession of magnetic field gradients (variation of fields in one direction in space) which, through the response of the spins, transmit the expected spatial information. MRI uses *linear magnetic field gradients* created by *gradient coils*. A gradient is said to be linear when its variation increases by the same value per unit of measurement in the chosen direction. These field gradients are superimposed linearly or antilinearly on the initial static magnetic field $\vec{H}_0$.

For example, consider a gradient application in the x direction (Figure 1.33). The $\vec{H}_0$ field increases linearly in the direction of gradient application with the addition of a $G_x \times x$ term. Since the Larmor frequency $\Omega_0 = \gamma_I H_0$, the precession frequency Ω of the proton spins will then depend on the strength of the overall external magnetic field. Each value of the applied magnetic field corresponds to a specific precession frequency in space. In this way, the magnetic field gradient enables the spatial localization of spins through frequency. In routine clinical MRI, *cross-sectional images* are generally taken.

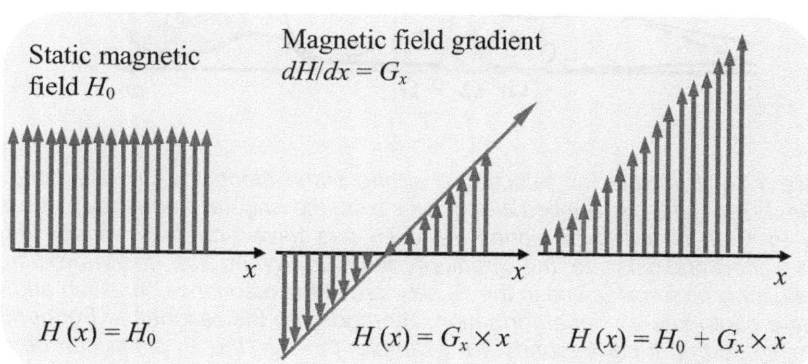

Figure 1.33. *Schematic illustration of the application of a magnetic field gradient in the Ox direction. For a color version of this figure, see www.iste.co.uk/sakho/nuclearphysics3.zip*

As an illustrative example, let us consider a slice plane perpendicular to the direction of the main magnetic field $\vec{H}_0$. The aim is to excite only the spins contained in this slice, so as to avoid any disturbance by the relaxation signal from spins in zones outside this slice during the signal recovery phase. To achieve this, a *slice selection gradient Gsc* is applied to select the desired slice plane, perpendicular to the direction in which the gradient is applied. This slice selection gradient *Gsc* superimposes a linear field on the $\vec{H}_0$ field, calibrated so that only one given slice plane is at resonance frequency during the 90° pulse (Figure 1.34).

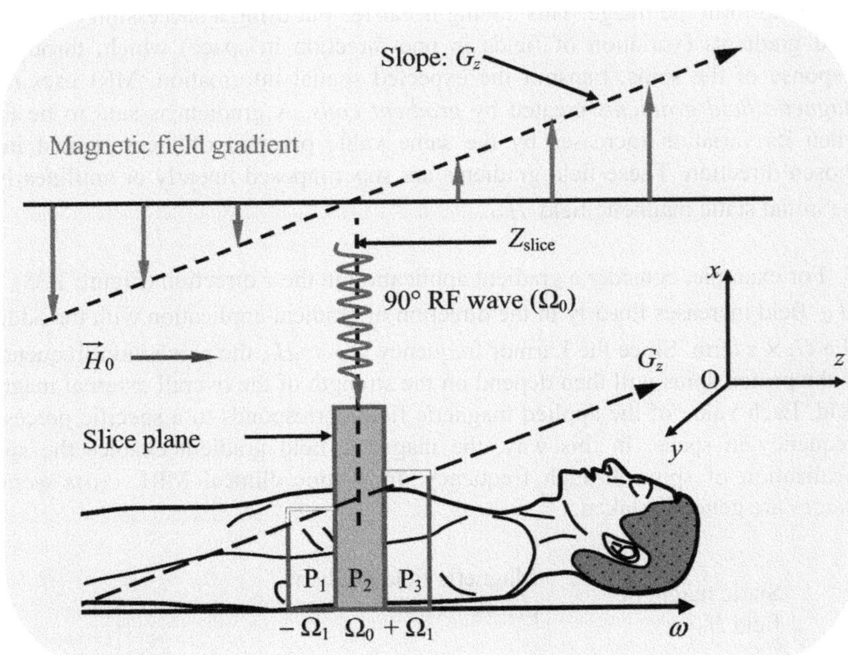

Figure 1.34. *Principle for selecting a slicing plane along the Oz direction. When a gradient Gsc (or G_Z) is applied along the z axis, the angular frequency increases from – Ω_1 to + Ω_1 in slices or planes P_1 to P_2 (we have arbitrarily chosen three slice planes) perpendicular to the gradient direction. When the resonance frequency $\Omega_r = \Omega_0$, only protons located in the P_2 slice are in the resonance condition and will tip at 90° and contribute to signal formation. By modifying the resonance frequency of the RF pulse so that it corresponds, for example, to – Ω_1, the P_1 plane can be selected. For a color version of this figure, see www.iste.co.uk/sakho/nuclearphysics3.zip*

Slice selection involves exciting all of the protons located in a layer of defined thickness. This selection makes it possible to isolate part of the tissue to be imaged.

If we denote the value of the gradient by G_z, the intensity of the magnetic field $H(z)$ linearly calibrated in the Oz axis is given by the formula:

$$H(z) = H_0 + G_z z \qquad [1.61]$$

In equation [1.54], z represents the position of the point on the Oz axis.

To image the slice at $z = z_{slice}$ corresponding to the P_2 plane (Figure 1.34), we apply excitation at the corresponding frequency, i.e. the slicing frequency ω_{slice} given by the equation:

$$\omega_{slice} = \gamma_1 H(z_{slice}) = \gamma_1 (H_0 + G_z z_{slice}) \qquad [1.62]$$

Under these conditions, only the spins located at position $z = z_{slice}$ will enter resonance. Bear in mind that RF excitation must be applied simultaneously with the slice selection gradient to obtain the desired signal.

NOTE.–

In Figure 1.34, we have arbitrarily chosen a slice perpendicular to the Oz direction. This means that the slice is not necessarily perpendicular to $\vec{H}_0$. It can be oriented differently in space by applying a combination of suitably calculated field gradients, so as to excite only the spins in the desired slice, following the principle outlined in the particular case of a slice perpendicular to the Oz direction.

1.4.9. *Frequency-encoding gradient and phase-encoding gradient*

In addition to the slice selection gradient, phase-encoding and frequency-encoding gradients are also applied to image selected tissues [ROQ 13, DUC 15, REN 16].

While the signal is being collected, let us apply a magnetic field gradient in a direction other than z (e.g. x). This gradient will modify the precession frequency of the protons. As the gradient is applied during signal collection, the signal will be made up of three frequencies rather than a single one (Figure 1.35). This is called *frequency coding* or *frequency-coding gradient*.

As soon as the gradient stops, the protons precess again according to Ω_0. The frequency-encoding gradient must therefore be applied during signal reception (or playback). For this reason, the gradient is also referred to as the *readout gradient*. Therefore, the pulsation of the signal emanating from each voxel in the chosen x-direction direction is directly proportional to its position on the x-axis. Let:

$$\omega(x) = \gamma_1 (H_0 + G_{readout} x) \qquad [1.63]$$

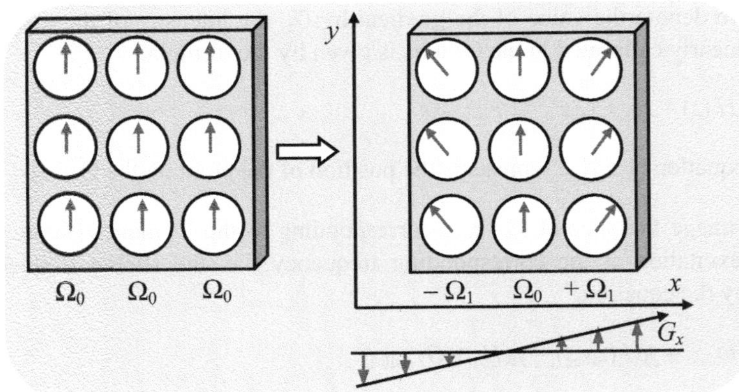

Figure 1.35. *Consequence of applying a Gcf (or G_x) frequency-coding gradient during signal reception. For a color version of this figure, see www.iste.co.uk/sakho/nuclearphysics3.zip*

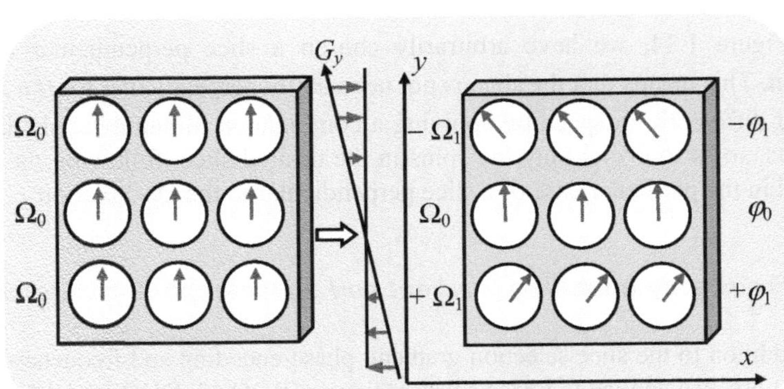

Figure 1.36. *Consequence of applying a Gcp or G_y phase-encoding gradient before signal reception. For a color version of this figure, see www.iste.co.uk/sakho/nuclearphysics3.zip*

Finally, a magnetic field gradient is applied in the remaining y direction before the signal is received. When the gradient is applied, spin precession is more or less rapid, depending on the position of the voxels in the y direction. On the one hand, the applied gradient will modify the proton precession frequency and, on the other hand, it will cause a parallel phase shift of the spins (Figure 1.36). When the gradient is stopped, the signal emanating from the various voxels is shifted in phase differently according to their y position. When the signal is recovered, each y-coordinate can be uniquely reassigned the corresponding signal according to its phase. This is known as *phase encoding* or G_{cp} *phase-encoding gradient.* It should also be noted that we have already mentioned sequences of phase encoding along y and frequency encoding along x. However, it is possible to perform these encodings in the same directions, but permuted.

An *NMR imaging sequence* is a chronological sequence of angular radiofrequency pulses θ and magnetic field gradients. The preceding coding sequences provide a chronogram of the basic spin echo sequence (Figure 1.37) or the basic gradient echo sequence (Figure 1.38).

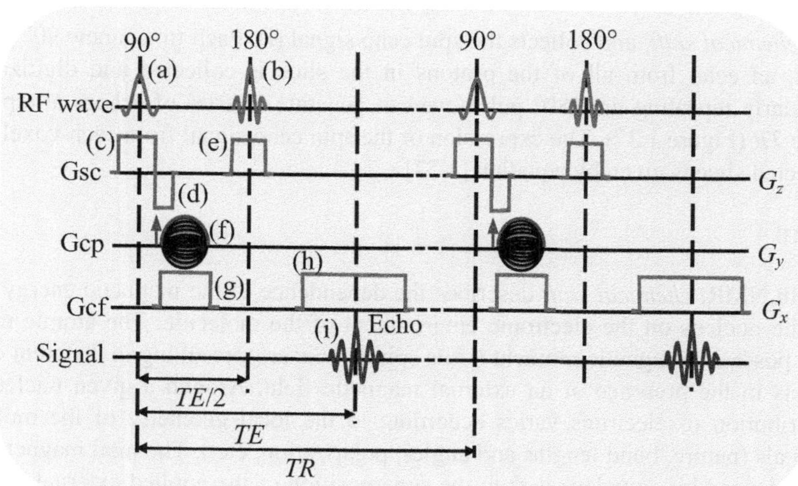

Figure 1.37. *Spin echo sequence timeline. For a color version of this figure, see www.iste.co.uk/sakho/nuclearphysics3.zip*

As explained above, the cut is selected by applying an RF wave pulse (a) and a slice selection gradient *Gsc* (c). During the application of the RF wave, the transverse component $\overrightarrow{M}_{xy}$ of the magnetization increases, reaching its maximum amplitude at $\theta = 90°$ (see Figure 1.16). When the excitation stops, the tilted magnetization returns to its thermal equilibrium state, under the influence of relaxation phenomena in T_1 and T_2. These phenomena are simultaneously reflected by a construction of the longitudinal component $\overrightarrow{M}_z$ according to a $1/T_1$ law [1.51] and an exponential destruction in $1/T_2$ [1.53] of the $\overrightarrow{M}_{xy}$ component in the transverse plane. Recall that amplitude decay is accompanied by a phase shift of the spins due to dipolar interactions (T_2) and H_0 field inhomogeneities (T_2*). This phase shift occurs throughout the time interval (0 – *TE*/2) after the 90° pulse. Next, the slice selection gradient is reversed (d) to rephase all of the spins of the slice excited at time *TE*/2 (Figure 1.37). The slice selection gradient (e) allows us to work on the slice selected in (c). The phase-encoding gradient (f) is applied in the same time interval (0 – *TE*/2) and at different amplitudes simultaneously with the application of a readout gradient (g) to rephase the spins. Beyond *TE*/2, a 180° rephasing pulse (h) is applied. This allows the spins to regain their phase coherence, and the transverse magnetization again reaches its amplitude maximum at time *TE* (see Figure 1.37). As we already know, this second 180° pulse overcomes the spin dephasing due to the constant inhomogeneities of the main magnetic field $\vec{H}_0$ and the *chemical shift*, and collects the spin echo signal (i). Each time a new 90° pulse is sent, an echo from all of the protons in the slice is collected and digitized. By regularly repeating the 180° pulse, we can generate a series of echoes at repetition time *TR* (Figure 1.37). The expression of the spin echo signal from each voxel of the selected slice is given by equation [1.57].

NOTE.–

In NMR, *chemical shift* describes the dependence of the magnetic energy levels of the nucleus on the electronic environment of the molecule. The atomic nucleus can possess a magnetic moment (if its spin is nonzero), resulting in different energy levels in the presence of an external magnetic field. Around a given nucleus, the distribution of electrons varies according to the local geometry of the molecular orbitals (nature, bond lengths and angles, polarization, etc.). The total magnetic field experienced by a nucleus is then the superposition of the applied external field and the local field induced by the presence and circulation of electrons. This has an impact on the energy levels and therefore the frequencies of nuclear magnetic resonance. These variations in the frequencies of the same type of nucleus are called "chemical shifts". The chemical shift is plotted against a reference frequency determined on a reference sample, often a molecule with a slightly distorted electronic distribution. The chemical shift is of great importance in NMR spectroscopy, particularly in organic chemistry, where it is one of the keys to determining the nature and structure of complex molecules.

Figure 1.38 shows a description of the chronogram of the gradient echo sequence. As previously mentioned, the gradient echo sequence does not have a 180° pulse, unlike the spin echo sequence. What is more, the first RF pulse is often shorter than 90° to reduce the time required for longitudinal regrowth (reduction in *TR* repetition time) and therefore image acquisition time.

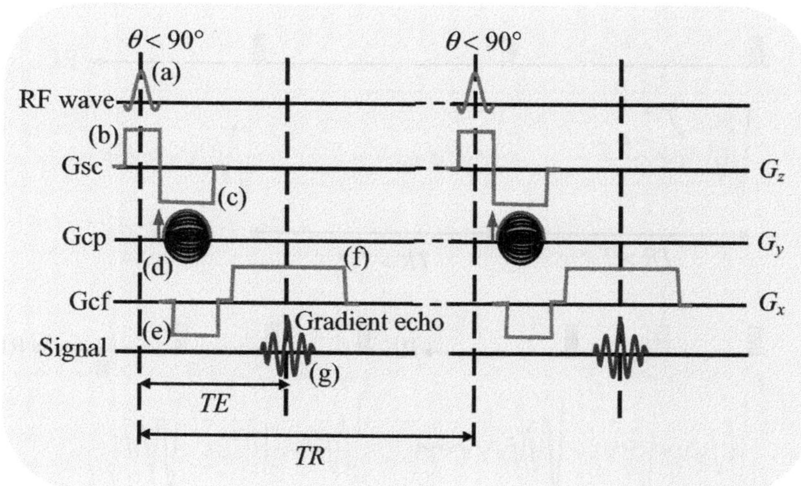

Figure 1.38. *Gradient echo sequence timeline. For a color version of this figure, see www.iste.co.uk/sakho/nuclearphysics3.zip*

To perform the gradient echo sequence, a slice selection pulse ($\theta < 90°$) (a) and a slice selection gradient (b) are applied. The selection gradient is then reversed (c) to rephase all proton spins in the selected excited slice. Next, a phase gradient (d) is applied in the same time interval simultaneously with the application of a readout gradient (e) to phase-shift the spins. Finally, a readout gradient (f) is applied both to rephase the spins in the middle of the echo and to acquire the gradient echo (g).

As shown in equation [1.59], the *gradient echo signal* Sge is a function of the tissue properties M_0, T_1 and T_2* and the acquisition parameters *TE*, *TR* and θ. Therefore, unlike the spin echo sequence, which is independent of T_2* (Sse depends on M_0, T_1, T_2, *TE* and *TR* (see [1.57])), the gradient echo signal is characterized by a residual transverse magnetization that is blurred at each cycle. This is due to the fact that in the *TE* time interval, the Sge signal decreases in T_2*, as the constant $\vec{H}_0$ field inhomogeneities of molecular origin (spin–spin relaxation) are not compensated for. This is a consequence of the absence of a 180° pulse to eliminate these inhomogeneities and therefore the access time T_2.

To eliminate residual magnetization during a gradient echo sequence, additional gradients are applied after the gradient echo has been acquired. These gradients are called *spoilers* (Figure 1.39).

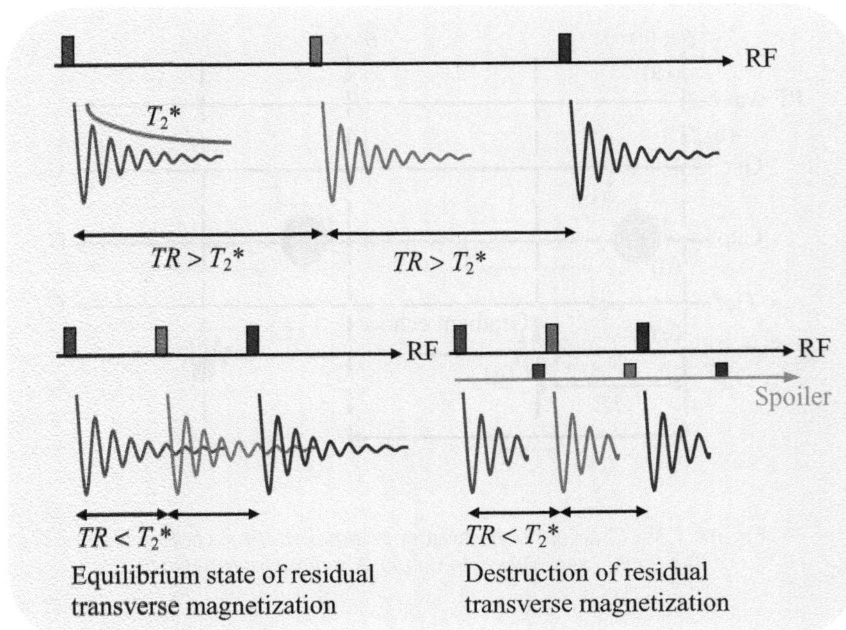

Figure 1.39. *Elimination of residual magnetization by applying additional gradients at the end of each gradient echo cycle. For a color version of this figure, see www.iste.co.uk/sakho/nuclearphysics3.zip*

1.4.10. *Fourier plane and MRI signal decoding*

In MRI, image acquisition is based on dual coding. On the one hand, we use frequency coding by applying a frequency-coding gradient in x, and on the other hand, phase coding by applying a phase-coding gradient in y. The data thus obtained is in the frequency domain, not the spatial domain. The information obtained is therefore a signal characterized by its frequency and/or phase. The space containing the raw data from MRI acquisition is called the *Fourier plane* or *k-space* (Figure 1.40).

To visualize a signal in terms of amplitude versus frequency, rather than amplitude versus time, we use a mathematical operation known as the *Fourier transform*. This transform extracts the individual frequencies of a signal (Figure 1.41).

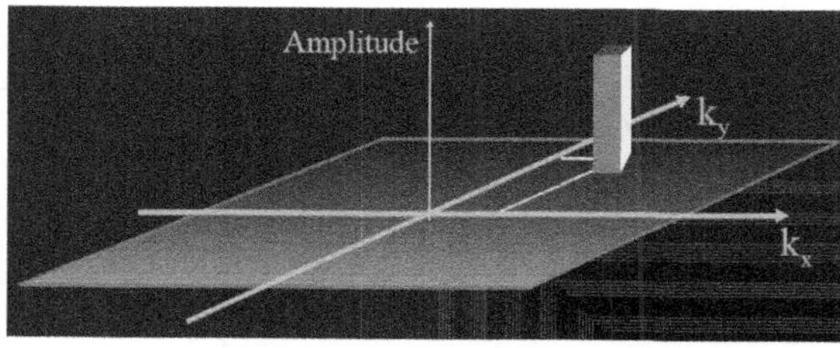

Figure 1.40. *Definition of the Fourier plane (k-space) (source: https://www.google.Bittoun.Imagerie+par+Resonance+Magnetique). For a color version of this figure, see www.iste.co.uk/sakho/nuclearphysics3.zip*

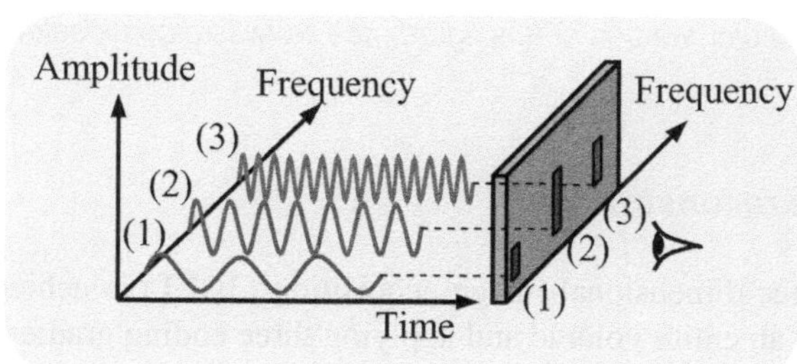

Figure 1.41. *Extraction of individual frequencies from a signal using the Fourier transform. For a color version of this figure, see www.iste.co.uk/sakho/nuclearphysics3.zip*

Under frequency- and phase-encoding conditions, the signal acquired in MRI is the Fourier transform of the slice image. Consequently, signal collection in *k*-space is necessarily followed by a decoding step. In this way, we move from the time domain to the frequency domain (Figure 1.41).

For any function not necessarily periodic, such as the FID signal (Figures 1.28 and 1.29) due to the decay of the transverse component of the macroscopic magnetization, the Fourier transform is visualized as a "continuous spectrum" represented by a continuous curve linking the different frequencies (Figure 1.42).

Since MRI is a digital imaging technique, it should be pointed out that to perform this operation with a computer, the analog signal received by the antenna must first be digitized using an *analog-to-digital converter* (ADC). The signal is sampled (a sequence of discrete signal values is taken at regular intervals), then the samples are quantized (a binary integer is assigned to each sample). The Fourier transform is then digital: *digital Fourier transform* (DFT). The DFT is very time-consuming to calculate, so to speed things up, we use a *fast Fourier transform* (FFT) [ROQ 13, REN 16].

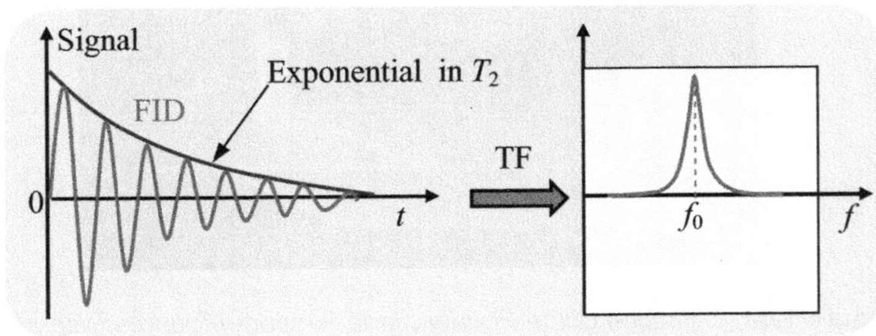

Figure 1.42. *Fourier transform (TF) of a non-periodic function (here FID signal). This signal decays according to the law $S(t) = S_0.e^{-t/T2} \times \cos(2\pi f_0 t + \varphi)$, the Larmor frequency $\Omega_0 = 2\pi f_0$. The Fourier transform gives a bell-shaped curve centered at $f_0 = \Omega_0/2\pi$. For a color version of this figure, see www.iste.co.uk/sakho/nuclearphysics3.zip*

1.4.11. *Image reconstruction*

In MRI, three-dimensional image acquisition (3DFT) is achieved by collecting the signal from an entire volume and applying three coding gradients along the three spatial directions x, y and z. As we know, the x direction corresponds to the frequency coding axis and y to the phase coding axis. By applying the different gradients, we obtain N_y signals acquired with different values of phase-encoding gradient G_{cp}. Each of these signals comprises N_x sampled in the presence of the readout gradient G_{lect} (G_{cf}). The Fourier plane is thus formed by a matrix of $N_x \times N_y$ points, made up of all of the signals obtained for a given slice.

For the x-axis, the spectral abscissa k_x is given by the equation:

$$k_x = \gamma_I \cdot G_{lect} \cdot t \quad [1.64a]$$

The same applies to the y-axis:

$$K_y = \gamma_I \cdot G_{cp} \cdot t \quad [1.64b]$$

In relations [1.64], the quantities G_{lect} and G_{cp} are also called readout gradient amplitude and phase gradient amplitude respectively, and $\gamma_I = \Omega_0/H_0$, the gyromagnetic ratio taken from relation [1.23].

The Fourier plane (see below) can be filled in a number of ways (spiral, horizontal line from the center, horizontal line from the edge, radial, etc.).

In MRI, the Fourier transform is used to pass from the frequency domain (understandable by the computer), to the spatial domain (the image, understandable by the human eye). So, to move from the spatial domain to the frequency domain, we apply the *direct Fourier transform* given by the following equation [EUS 15]:

$$TF\left[f(x,y)\right] = F(v_x,v_y) = \iint_{-\infty}^{+\infty} f(x,y)e^{-i2\pi(v_x x+v_y y)}dxdy \qquad [1.65a]$$

In equation [1.65a], $f(x, y)$ is a continuous function, x and y represent the spatial coordinates of a proton, and v_x and v_y denote the spectral coordinates: $v_x = \gamma\int G_x\, dt$ and $v_y = \gamma\int G_y\, dt$, with γ being the gyromagnetic ratio of the proton and G_x and G_y being the gradients of the applied magnetic field varying over time.

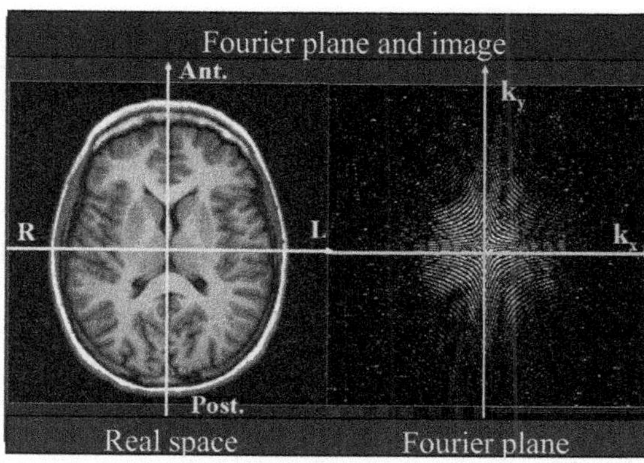

Figure 1.43. *Application of the direct Fourier transform to go from the image to the Fourier plane (source: https://www.google.Bittoun.Imagerie+par+Resonance+Magnetique). For a color version of this figure, see www.iste.co.uk/sakho/nuclearphysics3.zip*

In addition, to pass from the Fourier plane (frequency domain) to the image (spatial domain), we apply the *inverse Fourier transform* given by the equation [EUS 15]:

$$TF^{-1}\left[F(v_x,v_y)\right] = f(x,y) = \iint_{-\infty}^{+\infty} F(v_x,v_y)e^{i2\pi(v_x x+v_y y)}dv_x dv_y \qquad [1.65b]$$

Figure 1.44 shows the transition from the Fourier plane to the image by applying the inverse Fourier transform [1.65b].

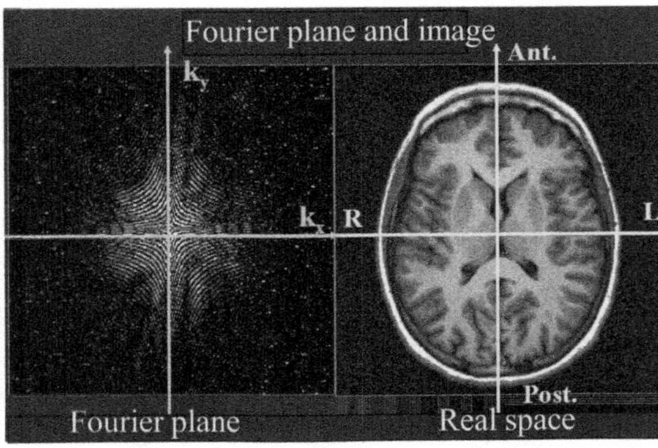

Figure 1.44. *Application of the inverse Fourier transform to go from Fourier plane to image (source: https://www.google.Bittoun.Imagerie+par+Resonance+Magnetique). For a color version of this figure, see www.iste.co.uk/sakho/nuclearphysics3.zip*

As shown in Figures 1.43 and 1.44, the constitution of the Fourier plane or *k*-space is such that the signal located at its center has a greater impact on the overall contrast of the final image ("global" image information) than the signal located at its periphery (details contained in the image). Note that this center corresponds to the low frequencies of the image, while the periphery is made up of the high frequencies.

Two-dimensional image acquisition involves the successive application of two coding gradients in x (frequency coding) and y (phase coding) directions. This double application is equivalent to a *double Fourier transform* (2DFT). Data acquisition in the Fourier plane is carried out line by line in a double progression, one horizontal and the other vertical.

– Horizontal progression in the x direction consists of scanning or reading the same line corresponding to the frequency coding by the readout gradient Gef (Frequency-Coding Gradient);

– Vertical progression in the y direction consists of moving from one line to the next by "incrementing" (increase) the Gcp phase-coding gradient to successively obtain the n of the Fourier plane (Figure 1.45).

As we know, in MRI, a complete "image" cycle corresponds to 128 or 256 lines. The repetition time TR (time between two cycles or two 90° pulses) corresponds to the passage from one line to another. A line of the Fourier plane is acquired, during a cycle, at the end of the echo time TE (Figure 1.45). The image is obtained from *k*-space by *inverse Fourier transform*, using a signal processing algorithm that performs this transformation very rapidly.

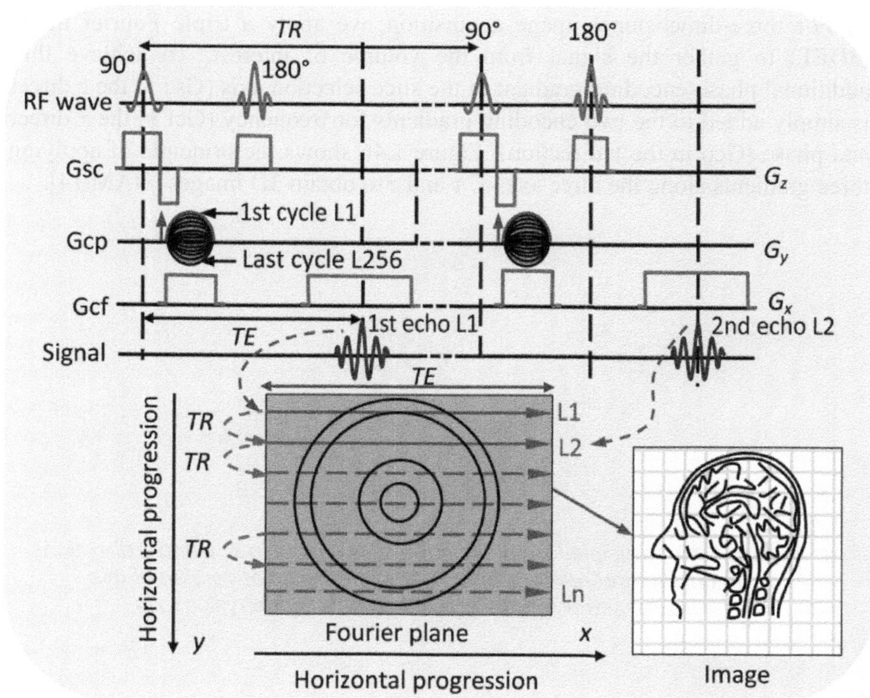

Figure 1.45. *Principle of data acquisition in the Fourier plane. The image is obtained from k-space by inverse Fourier transform. For a color version of this figure, see www.iste.co.uk/sakho/nuclearphysics3.zip*

The *acquisition time* T_{ac} required to obtain an image depends on several parameters. It is given by the equation [EUS 15, REN 16]:

$$T_{ac} = TR \times N_p \times N_{ex} \qquad [1.66]$$

where:

– TR is the repetition time;

– N_p denotes the *number of rows* in the matrix $N_x \times N_y$ (number of times the phase-encoding gradient is incremented to fill a row in the Fourier plane);

– N_{ex} represents the *number of excitations* or accumulations (number of measurements of the same line).

Note that the dependence of acquisition time on the number of N_{ex} excitations is justified by the desire to improve image quality (signal-to-noise ratio S/N). To achieve this, it is necessary to pass over each line several times.

For three-dimensional image acquisition, we apply a triple Fourier transform (3DFT) to gather the signal from the volume of interest. To achieve this, an additional phase-encoding gradient in the slice selection axis (Gsc in the z direction) is simply added to the two encoding gradients for frequency (Gcf in the x direction) and phase (Gcp in the y direction). Figure 1.46 shows the principle of applying the three gradients along the three axes x, y and z to obtain 3D images [HAM 14].

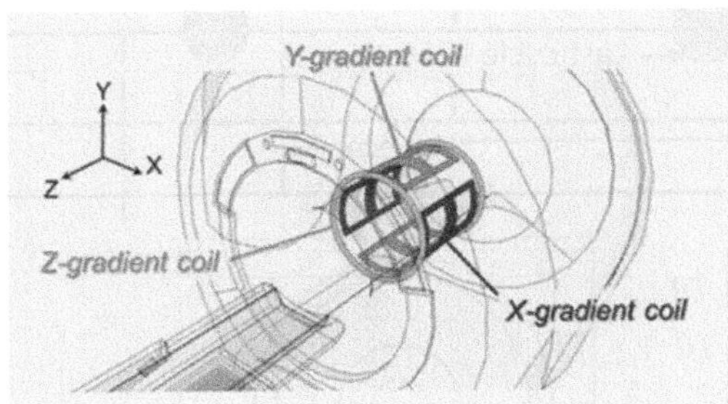

Figure 1.46. *Principle for applying three gradients in x, y and z directions to obtain three-dimensional images. For a color version of this figure, see www.iste.co.uk/sakho/nuclearphysics3.zip*

NOTE.–

It is the vibration of the magnetic field gradients that induces the powerful noises emitted by the MRI scanner. It should be noted that the noise is also the result of noise coming from the patient (thermal agitation of protons causing spurious emissions). The signal-to-noise ratio is a function of the average signal intensity in relation to the amount of noise variation. This ratio depends on:

– non-modifiable factors: field strength, RF wave sequence, tissue characteristics;

– modifiable factors: antenna used;

– sequence parameters: voxel size (spatial resolution), number of excitations, reception bandwidth.

If we make several measurements for the same slice, then for each measurement:

– the signal will be identical each time;

– the noise will be different (because it is random).

By cumulating several measurements for a single slice, the signal increases proportionally to the number of measurements, while the noise increases proportionally to the square root of the number of measurements. So, by averaging several measurements for a single slice, the average signal obtained remains constant, while the variability of the noise decreases. This justifies the importance of the number of excitations in expressing the acquisition time [1.66].

NOTE.–

Edouard Mills Purcell (1912–1997) was an American physicist. He is best known for his work on nuclear magnetic resonance (NMR) in liquids and solids. He shared with Félix Bloch the 1952 Nobel Prize in Physics for their development of new methods for fine nuclear magnetic measurements and the discoveries that followed. Purcell also made discoveries in astronomy. He was the first to detect radio emissions from galactic hydrogen, providing the first "image" of the spiral arms of the Milky Way.

Félix Bloch (1905–1983) was an American physicist. He is best known for his work on nuclear magnetic resonance, for which he shared the 1952 Nobel Prize in Physics with Purcell. In 1946, Bloch developed a set of macroscopic equations known as the Bloch equations, used in NMR, MRI and EPR (electron paramagnetic resonance). These equations, sometimes called the equations of motion of nuclear magnetization, are used to calculate nuclear magnetization $\mathbf{M} = (M_x, M_y, M_z)$ as a function of time when relaxation times T_1 and T_2 are present.

Paul Christian Lauterbur (1929–2007), an American chemist, and Sir Peter Mansfield (1933–2017), a British physicist, are best known for their work on magnetic resonance imaging. They were cowinners of the 2003 Nobel Prize in Physiology or Medicine for their development of MRI.

1.5. Ultrasound imaging

1.5.1. *Ultrasound principle*

A transducer emits high-frequency *ultrasonic waves* that are inaudible to the human ear (the high-pitched sound could be heard by dogs and dolphins). Remember that the frequencies of audible sound waves vary between 20 Hz and 20,000 Hz. Ultrasound at frequencies $f > 20$ kHz travels through air with the same speed as sound, equal to 340 m · s $^{-1}$. In a gas such as dihydrogen H_2, sound propagates at 1,300 m · s $^{-1}$. This speed rises to 1,500 m · s $^{-1}$ in water, the order of magnitude of the propagation speed of sound waves in the body. Tissues present a resistance to the passage of *ultrasound*. This resistance depends on the tissue traversed, its elasticity and density [POI 19]. The principle of ultrasound is shown in Figure 1.47.

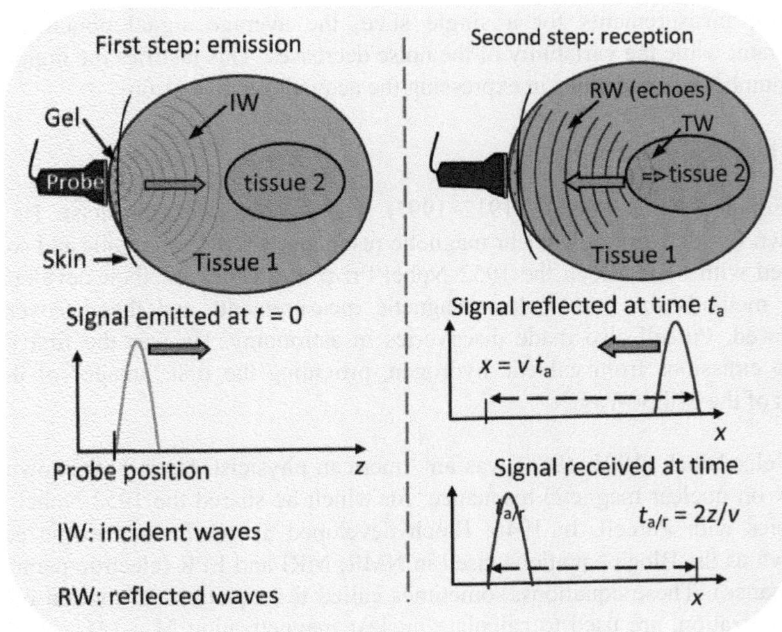

Figure 1.47. *Ultrasound principle. The echo signal is received after a round trip $t_{a/r} = 2x/v$. For a color version of this figure, see www.iste.co.uk/sakho/nuclearphysics3.zip*

A probe containing an ultrasound transmitter and receiver is moved over the patient's skin (first step). Ultrasound then arrives at the interface between two tissues numbered 1 and 2. Depending on the tissue encountered, some of the ultrasound waves are transmitted (OT) and some are reflected (OR). The higher the impedance difference between tissues 1 and 2, the greater the reflection. The reflected ultrasound waves, known as echoes, are therefore used to identify, locate and characterize the various tissues they encounter. A receiver on the probe receives the echoes (waves reflected by the various tissues traversed) and transmits them to a computer system that reconstitutes a flat or 3D *ultrasound image*.

NOTE.–

It should be noted that the echoes received during an examination are less and less powerful due to the attenuation phenomena that the wave undergoes during its outward and return journey. Signal amplifiers are used to increase the signal over time (and therefore the depth of the echo) flows from the moment of transmission. This is called time gain control (TGC) (see: https://www.cnp-mn.fr/les-fondamentaux-chapitre-7-echographie/, 2024).

VOCABULARY CORNER.–

Transducer: a device that transforms one physical quantity into another.

Ultrasonic transducer: a device that uses the piezoelectric effect to convert electricity into mechanical vibration and form ultrasonic waves.

Piezoelectricity: the property of certain materials to become electrically polarized under mechanical stress and vice versa to deform when an electric field is applied to them. The piezoelectric effect is used to convert mechanical energy into electrical energy and vice versa.

1.5.2. *Different elements of an ultrasound scan*

To perform an ultrasound scan, gel is placed between the probe and the skin to facilitate penetration of the ultrasound waves into the body. Many organs are therefore visualized, whether superficial (breasts, testicles, thyroid) or deep (kidneys, liver, pancreas, uterus, etc.), as well as muscles and tendons. When *gel* is not used during an *ultrasound examination*, a thin layer of air is present between the probe and the patient's skin. This results in poor image visibility. Ultrasound gel is therefore essential for obtaining a high-quality image. This gel is composed of purified water, gelling agents, preservatives and sometimes colorants.

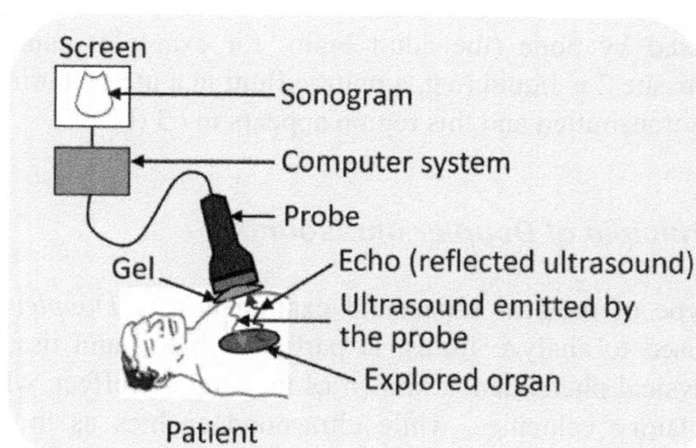

Figure 1.48. *Various elements involved in ultrasound scanning. For a color version of this figure, see www.iste.co.uk/sakho/nuclearphysics3.zip*

Knowing the propagation speed of the ultrasound wave in the tissue, and measuring the time $t_{a/r}$ between the moment when the signal is emitted and the moment when it is received, we can determine the distance at which tissue 2 is located (first step, Figure 1.47). All of the elements required to obtain an ultrasound image are shown in Figure 1.48.

Ultrasound is a very useful technique for monitoring pregnancies (obstetrics), cardiology (despite interference from the ribs), gastroenterology (liver, bile, kidneys), urology (genital tract, bladder), etc. However, due to ultrasound opacity in air-containing areas, the lungs, digestive tract and bones cannot be analyzed by sonography or ultrasonography.

NOTE.– In relation to [POI 19]

i) The probe is the basic element of ultrasound. It is made of piezoelectric ceramic. It has the dual role of emitter and receiver of ultrasound. As a transmitter, it is subjected to electrical impulses that generate ultrasound. As a receiver, it is referred to as an ultrasound transducer, and the ultrasound wave induces an electrical impulse that gives rise to the ultrasound image. Typically, a thousand ultrasound transducers make up a probe. Finally, each probe has its own operating frequency, higher or lower depending on the application. Typically, a 1.5 MHz probe will enable acquisitions of the abdomen with a spatial resolution of the order of a mm. Probes operating at 7 MHz will be used for superficial imaging of veins or arteries, with a spatial resolution of the order of a tenth of a mm.

ii) If tissue 2 (Figure 1.47) is bone, all of the signal is reflected, and the bone appears hypersignal (white). The transmitted wave is practically nil, which is why regions protected by bone (the adult brain, for example) cannot be imaged by ultrasound. If tissue 2 is liquid (e.g. amniotic fluid in a uterus during pregnancy), the entire signal is transmitted and this region appears in 63 (black).

1.5.3. *The principle of Doppler ultrasound*

Another type of medical ultrasound examination is *Doppler ultrasound*. This technique is used to analyze fluids, in particular blood and tissue irrigation. It is based on a physical phenomenon known as the Doppler effect, which enables us to quantify circulatory velocities, while ultrasound enables us to visualize vascular structures (blood vessel network). The Doppler effect is the variation in frequency of a light signal when the source and emitter are in relative motion (see Volume 2, Chapter 1, section 1.1). In the case of the medical examinations that interest us in this book, the ultrasound probe and the blood are in relative motion. This justifies the use of the Doppler ultrasound technique to quantify blood flow velocities in vessels.

The Doppler ultrasound examination lasts between 10 and 20 minutes. The procedure is similar to a follow-up ultrasound examination. The gel used is spread on the skin to ensure good transmission of the ultrasound emitted by a probe applied opposite the structure under examination. The ultrasound wave propagates through the tissues and is reflected as an echo by the various organs it encounters. The signal is analyzed and transformed into a sound, a curve or a color reflecting blood flow speeds and direction. The Doppler system is integrated into the ultrasound machine. Combined with ultrasound, it provides information on vessel shape. It can be used on vessels in the limbs, neck and abdomen. It mainly looks for disturbances in blood flow that may be related to an obstacle or narrowing of the vessel. It can also be used to detect a blocked *clot* in a vein. Figure 1.49 illustrates the principle of Doppler ultrasonography for the study of blood flow in a vessel such as an artery or vein.

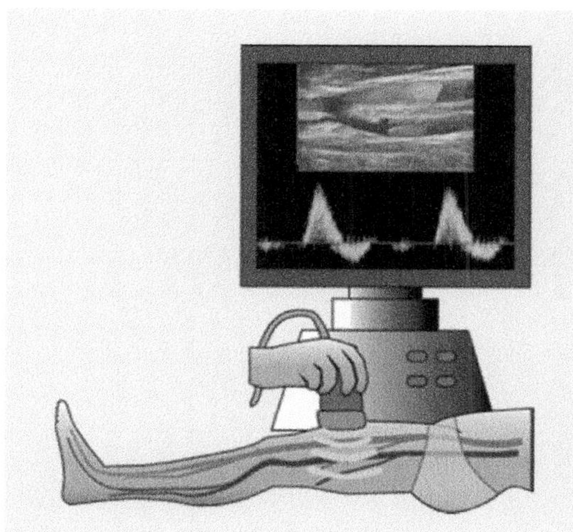

Figure 1.49. *Principle of Doppler ultrasound for exploring blood flow in a vessel (source: https://www.imagir.org/vos-examens/examens-echographiques/doppler/). For a color version of this figure, see www.iste.co.uk/sakho/nuclearphysics3.zip*

In conclusion, morphological imaging (radiology, ultrasound and MRI) is still limited to the anatomical analysis of results. This imaging can be used to identify organs, check their conformation and, if necessary, monitor changes in shape over time, particularly growth or reconstitution. Morphological imaging, on the contrary, does not enable doctors to determine whether a tissue under study is functioning normally. By contrast, nuclear medicine imaging enables both morphological analysis of organs and functional analysis of tissues. The precision of the measurements will remain to the advantage of the first three methods (radiology, ultrasound, MRI). Doctors are well aware of this, and also know that the four methods are complementary, each providing different levels of information. The recent development of mixed tools combining these morphological imaging technologies (radiology, ultrasound, MRI) and functional imaging (SPECT, PET) meets this need [DIL 92]. It should be noted that a patient who has just undergone an MRI examination is not magnetized.

2

Technetized Radiopharmaceuticals and Radiothallium-201 used in Nuclear Medicine

General objective	
Understand the properties and uses of metastable technetium-99-labeled radiopharmaceuticals and radiothallium-201 in nuclear medicine.	
Specific objectives	
Make the link between marker and radiotracer or radiopharmaceutical;	Understand the principle of ^{99m}Tc-MAA (^{99m}Tc-Macroalbumin Aggregate) lung perfusion scintigraphy and the dosimetry used;
Understand the principle of Anger gamma cameras (γ-camera);	Explain the mechanism of ^{99m}Tc-MAA clearance in the body after IV injection;
Understand the usefulness of the latest generation of CZT (Cadmium Zinc Tellurium) cameras;	Understand the radiothallium-201 production chain;
Understand the value of the two main radiotracers ^{99m}Tc and ^{201}Tl n myocardial scintigraphy;	Learn about the main emissions of thallium-201;
Describe the ^{99}Mo-^{99m}Tc generator production line;	Understand the principle of ^{201}Tl myocardial perfusion scintigraphy and the dosimetry used;
Describe the process for preparing metastable technetium-99 from the ^{99}Mo-^{99m}Tc generator;	Explain the mechanism of radiothallium-201 clearance in the body after IV injection;
Define the concepts of effective half-life, biological half-life and physical half-life;	Understand the properties of the two radiopharmaceuticals ^{99m}Tc-HMPAO (^{99m}Tc-HexaMethylPropylene-Amine-Oxime) and ^{99m}Tc-ECD (^{99m}Tc-Ethylenediyl bis-L-Cysteine-Diethyl-ester) used in cerebral perfusion scintigraphy;
Use the relationship between effective half-life, biological half-life and physical half-life;	Understand the main characteristics of ^{99m}Tc-HMPAO and ^{99m}Tc-ECD radiopharmaceuticals;

Understand the main characteristics of ^{99m}Tc-sestamibi or MIBI ^{99m}Tc (2-Methoxy-IsoButyl Isonitrile-^{99m}Tc), ^{99m}Tc and ^{201}Tl tetrofosmin tracers;	Understand the principle of ^{99m}Tc-HMPAO brain perfusion scintigraphy and the dosimetry used;
Understand the advantages of thallium-201-based tracers;	Understand the principle of ^{99m}Tc-ECD brain perfusion scintigraphy and the dosimetry used;
Explain the phenomenon of thallium-201 redistribution in the body;	Understand brain perfusion scintigraphy examination procedures;
Understand the distribution mechanism of MIBI ^{99m}Tc after intravenous (IV) injection;	Understand the properties of the radiopharmaceutical ^{99m}Tc-MDP (^{99m}Tc-Methyl DiphosPhonate);
Understand the principle of myocardial scintigraphy;	Understand the principle and procedure of bone scintigraphy;
Justify the preference for MIBI ^{99m}Tc over 201Tl in terms of dosimetry and the γ-camera used;	Understand the properties of the radiopharmaceuticals ^{99m}Tc-DMSA (DiMercapto Succinic Acid), ^{99m}Tc-MAG$_3$ (*Mercapto Acetyl triglycine*) and ^{99m}Tc-DTPA (^{99m}Tc-Diethylene Triamine Penta Acetic) used in renal scintigraphy;
Understand the properties of MIBI ^{99m}Tc;	Understand the main characteristics and properties of the radiotracers ^{99m}Tc-DMSA, ^{99m}Tc-MAG$_3$ and ^{99m}Tc-DTPA;
Explain the principle of ^{99m}Tc-MIBI myocardial scintigraphy and the dosimetry used;	Understand the principle of ^{99m}Tc-DMSA static renal scintigraphy and the dosimetry used;
Understand the principle of ^{99m}Tc tetrofosmin myocardial scintigraphy and the dosimetry used;	Understand the principle of ^{99m}Tc-DTPA dynamic renal scintigraphy and the dosimetry used;
Explain the mechanism of ^{99m}Tc-MIBI clearance in the body;	Understand the procedure for dynamic ^{99m}Tc-TDPA and ^{99m}Tc-MAG$_3$ scintigraphy;
Explain the principle of ^{99m}Tc Technegas® lung ventilation scintigraphy and the dosimetry used;	Understand the clearance mechanism of ^{99m}Tc-DMSA, ^{99m}Tc-MAG$_3$ and ^{99m}Tc-DTPA radiopharmaceuticals in the body after IV injection;
Explain the clearance mechanism of Technegas® in the body after inhalation;	Understand the principle of digestive scintigraphy or gastric emptying scintigraphy;
Explain the principle of ^{99m}Tc-DTPA (^{99m}Tc-diethylenetriamine pentaacetic acid) lung ventilation scintigraphy and the dosimetry used;	Know the nature of the metastable technetium-99-labeled consensus standardized meal;
Understand of the main characteristics and properties of ^{99m}Tc-DTPA;	Understand the procedure for digestive scintigraphy.
Prerequisites	
Beta decay;	Isomeric transition;
Electronic capture (EC) process;	Nuclear transmutation reactions;
Auger effect;	Anatomies of the heart, lungs, brain, kidneys and stomach.
Gamma de-excitation;	

2.1. Metastable technetium-99 and thallium-201 radiotracers

2.1.1. *Overview*

As we pointed out in section 4.4 of Chapter 4 of Volume 2, radiopharmaceuticals used in medical imaging may be isolated radionuclides or may consist of a carrier and a radionuclide. Examples of isolated radionuclides are iodine-123, iodine-125, iodine-131 (see Chapter 4, Volume 2), thallium-201 and metastable krypton-81. Clinically, thallium-201 chloride is used without a vector for IV myocardial scans. Similarly, metastable krypton-81 is used without a carrier and in aerosol form for lung ventilation scans (see section 2.4 of the present chapter).

Myocardial scintigraphy mainly uses two radiotracers: thallium-201 (^{201}Tl) and metastable technetium-99 (^{99m}Tc), both of which bind to the myocardium [MEN 06, VER 11, HAN 16].

Technetiated tracers (sestamibi and tetrofosmin) passively enter the cell, but bind to the mitochondrial membranes of living cells. Once inside the cell, the product remains there and there is no redistribution. Passage under the γ-camera can be immediate or delayed, which is usually the case to avoid visualizing the product in the digestive tract (natural elimination). The 140 keV radiation emitted is better suited to the cameras used, and attenuation phenomena are less numerous. The half-life of the product is 6 h, and dosimetry is more favorable [MAR 15]. Thallium-201 emits low-energy gamma rays (69 and 80 keV) and has a half-life of 73 h. Due to the existence of so-called internal conversion decay, its dosimetry is not favorable. Table 2.1 shows the main characteristics of MIBI ^{99m}Tc, Tetrofosmin ^{99m}Tc and ^{201}Tl tracers used in myocardial scintigraphy.

Tracer	Period (in h)	Energy (in keV)	Image quality	Dosimetry	Redistribution
^{201}Tl	73	80 (weak)	Good	35 mSv (important)	Yes
MIBI ^{99m}Tc Tetrofosmin ^{99m}Tc	6	140 (adequate)	Very good	9 mSv (weak)	No

Table 2.1. *Main characteristics of MIBI ^{99m}Tc, Tetrofosmin ^{99m}Tc and ^{201}T tracers*

NOTE.– Redistribution phenomenon

Thallium-201, an isotope with a structure close to that of potassium, enters cells via the sodium/potassium pumps, by means of an active transport mechanism.

Importantly, thallium does not remain in the cell, but can leave it via the Na/K pump system. This is the so-called redistribution phenomenon, where thallium was expected to be redistributed on images in early publications. This also means that the patient must be placed under the camera immediately after the stress test, otherwise the images will not accurately reflect myocardial perfusion during exercise [MAR 15].

Technetium (^{99m}Tc)-sestamibi is a cationic complex that passively diffuses across capillary and cell membranes. Within the cell, it is captured and retained in mitochondria, and its retention reflects myocardial cell viability. After intravenous injection, it is distributed throughout the myocardium according to myocardial perfusion and viability. Myocardial uptake, which depends on coronary flow, is 1.5% of the activity injected during exercise and 1.2% of the activity injected at rest. However, irreversibly damaged cells no longer take up technetium (^{99m}Tc)-sestamibi. Hypoxia reduces the rate of myocardial extraction. Its redistribution is insignificant, and separate injections are therefore required for examinations after pacing and at rest [MIN 17].

VOCABULARY CORNER.–

Capillaries: the finest and smallest blood vessels that exist in vertebrates. Blood vessels are the channels through which blood circulates (artery, vein or small capillary vessel).

Mitochondria: elements of the cytoplasm of the animal or plant cell. Their essential role is to ensure oxidation, cellular respiration, the storage of energy by the cell and the storage of certain substances such as calcium, sodium and potassium ions.

2.1.2. *Dosimetry and gamma cameras*

In myocardial scintigraphy, radiotracers have the ability to bind specifically to the target organ in question. The γ-radiation emitted by these radiotracers is captured by a conventional γ-camera (so called to distinguish them from the latest-generation γ-cameras of the CZT type). This detects the signal from the organ under study and reconstructs a digitized static or dynamic image. In general, dosimetry depends on the activity injected and the type of γ-camera used. With the advent of the latest generation of CZT (Cadmium Zinc Tellurium) cameras, this dosimetry is more reduced compared with conventional γ-cameras, as shown in the data presented in Table 2.2 [MAR 15].

Isotope	Thallium 201 (^{201}Tl)		MIBI ^{99m}Tc	
Camera	**Gamma camera**	**CTZ camera**	**Gamma camera**	**CTZ camera**
Two passages	35 mSv	26 mSv	9 mSv	6 mSv
One passage	26 mSv	18 mSv	2.5 mSv	1.5 mSv

Table 2.2. *Dosimetry according to isotopes and cameras used. For a color version of this table, see www.iste.co.uk/sakho/nuclearphysics3.zip*

The data given in Tables 2.1 and 2.2, concerning image quality and absorbed dose in particular, justify the preference for ^{99m}Tc over ^{201}Tl in myocardial scintigraphy. Furthermore, as shown in Table 2.2, the type of γ-camera used influences the dosimetry to be applied.

The latest-generation CZT γ-camera is a solid-state camera that replaces conventional γ-cameras. It has proven its worth in cardiac scintigraphy. It enables faster, more comfortable and less irradiating scintigraphy for a more reliable diagnosis. In fact, the examination is three times faster and six to eight times more sensitive than a conventional γ-camera. Examination time is reduced to 5 min, and a complete scan of myocardial perfusion during exercise and at rest can be performed in less than 2 hours. Patient positioning and comfort are also greatly improved [MUR 11].

Semiconductor γ-cameras are based on the principle of direct conversion of the incident gamma photon (γ) into electron–hole pairs (an electrical signal) proportional to its energy. Under the action of an electric field applied between the electrodes, the charges move: electrons to the anode and holes to the cathode. The electrical output signal from each electrode then has an amplitude proportional to the energy of the incident γ-photon. In contrast to the indirect conversion principle of conventional γ-cameras (conversion of the incident gamma photon into light photons and then into an electrical signal), the interaction of a gamma photon of energy 140 kiloelectron volts (keV) in a CZT material produces around 30,000 electron–hole pairs, i.e. some 20 times more than with a scintillating NaI crystal [BLA 17].

Note that the equivalent dose administered to a patient during a scintigraphic examination is measured in sieverts. There are therefore two fundamental units in radiation protection: the gray and the sievert.

The *gray* (symbol Gy) is the unit of absorbed dose. The absorbed dose, or D-dose, is the physical measure of the energy of ionizing radiation absorbed by a living organism. Therefore, one gray represents the energy of ionizing radiation delivering an energy of one joule to a homogeneous biological medium with a mass of one kilogram. By definition [SAK 17]:

$$D = \Delta W / \Delta m \quad [2.1]$$

where:

– ΔW: the absorbed radiation energy (ΔW in J);

– Δm: the irradiated mass (Δm in kg);

– D: the absorbed dose, expressed in gray (Gy): 1 Gy = 1 J/kg (or 1 J · kg^{-1}).

It should be noted that in organic tissues and water, an absorbed dose of 1 Gy only causes a temperature rise of 0.0024°C. However, this dose is sufficient to destroy many vital cells, as the energy deposited is highly localized and concentrated along the path of the ionizing radiation.

The sievert (symbol Sv) is the equivalent dose. The equivalent dose H is the absorbed dose D_R by an organism weighted by a factor w_R taking into account the nature of the radiation R. By definition [SAK 17]:

$$H = w_R . D_R \quad [2.2]$$

where:

– w_R: the unitless radiation-dependent weighting factor R; D_R: the absorbed radiation dose R, expressed in gray.

It should be noted that the equivalent dose expresses the fact that, for the same absorbed doses, the biological effects differ depending on whether they are photons, various particles (α, β, protons, etc.) or neutrons. The lethal dose is around 5 sieverts (Sv). The weighting factor for R is assessed on the basis of its (*linear energy transfer*) in a medium. LET is the energy transferred by a particle per unit length of material traversed. LET is generally expressed in keV/μm (or keV · μm^{-1}). The weighting factor is set by the International Commission on Radiological Protection (ICRP). Table 2.3 gives some w_R for different tissues [SAK 17].

Radiation type	w_R
X-ray and γ-radiation	1
Electrons, positrons, muons	1
Neutrons: $E < 10$ keV or $E > 20$ MeV	5
Neutrons: 10–100 keV or 2–20 MeV	10
Neutrons: 0.1–2 MeV	20
α-particles and heavy ions	20

Table 2.3. *w_R weighting factor values for different tissues*

In dosimetry, the equivalent dose rate is defined as the ratio of the equivalent dose to the duration of irradiation. It is expressed in Sv/s or Sv/h, or very rarely in Sv/year. In addition, until 1986, the rad was used to express the unit of absorbed dose: 1 rad = 0.01 Gy = 10 mGy. Rad stands for ***r**adiation **a**bsorbed **d**ose*. Similarly, until 1979, rem was used to express the unit of equivalent dose: 1 rem = 0.01 Sv = 10 mSv [TAI 08]. Rem stands for ***r**oentgen **e**quivalent **m**an*.

The gray is used in particular to assess the deterministic effects (see the note below) of high doses of radiation on humans, while the sievert is used to assess the stochastic effects of low doses. Deterministic effects result from irreparable damage to DNA caused by a high dose of radiation. Stochastic effects, on the contrary, relate to the effects of low doses of toxic substances or the health-damaging effects of ionizing radiation.

VOCABULARY CORNER.–

– *Dose rate*: the amount of radiation delivered over a given time. A dose delivered slowly (low-dose-rate) does not have the same effects as the same dose delivered more rapidly (high-dose-rate).

– *Dosimetry*: the calculation of radiation doses to be applied to the area to be treated and the duration of treatment during radiotherapy.

– *Lymph node*: small bulges along the lymphatic vessels. Often arranged in chains or clusters, lymph nodes are either superficial (in the neck, armpit, groin) or deep (in the abdomen, thorax). They play an essential role in protecting the body against infection or cancerous cells. They normally measure less than a centimeter in diameter. If their size is abnormal, they are called adenopathies. An enlarged lymph node can be linked to anything other than cancer.

– *Gland*: an organ whose function is to produce certain substances which it then releases, either into the blood (endocrine gland) or outside the body (exocrine gland).

– *Lymphatic vessels*: channels through which lymph circulates. The lymphatic vessels connect the nodes together to form the lymphatic system, involved in the body's defense. Lymph is a whitish or yellowish biological fluid which circulates in the lymphatic vessels. Its composition is similar to that of blood plasma, of which it is only a filtrate. It contains white blood cells, notably lymphocytes. Lymph collects certain wastes, bacteria and damaged cells from inside the body's tissues so that they can be flushed out of the body or destroyed.

BIOGRAPHY CORNER.–

Rolf Sievert (1896–1966) was a Swedish physicist. He is known for his many contributions to the development of radiation protection. In 1929, he initiated the creation of the International Commission on Radiological Protection (ICRP), which he headed from 1956 to 1962, and the International Commission on Radiation Units and Measurements (ICRU). In the 1930s, Sievert began work on the biological effects of ionizing radiation. His most important contribution to clinical physics was to develop the basis for calculating the absorbed doses needed to treat tumors. In his honor, Resolution 5 of the XVIth International Conference on Weights and Measures (CGPM) in 1979 defined the sievert (Sv) within the SI system, as the unit of equivalent dose of ionizing radiation.

NOTE.–

The effects of ionizing radiation depend on its nature, the dose absorbed and the organ affected. They are classified into two categories: effects with deterministic or non-stochastic incidences and effects with random or stochastic incidences.

Deterministic effects occur with certainty. For high doses (greater than several grays), they are observed in all exposed subjects. These are therefore threshold effects. They generally appear early, with latency times of between a few days and a few months. Their severity increases with the dose absorbed.

We begin to observe certain deterministic effects around 0.3 Gy for partial exposures. On the contrary, for low doses, below a threshold value depending essentially on the type of biological effect, no effect is detectable. For overall exposure, we take the reference value of 0.5 Gy as the threshold value.

From a certain absorbed dose (around 2 Gy), for overall exposure, there is a risk of death. We call the 50% lethal dose (LD50), the absorbed dose, for the entire body (therefore overall exposure), for which the probability of dying 60 days after exposure, without medical treatment, is 50%. It is equal to 4.5 Gy (or between 4 and 4.5 Sv). Generally, deterministic effects occur in the case of a single exposure at a high-dose-rate. These are therefore often in the case of accidental situations.

Exposure to higher or lower doses of ionizing radiation can have long-term effects in the form of cancer or genetic mutations affecting offspring. An exposure can be external or internal. For example, radiation of cosmic origin or during a medical examination: radiodiagnosis (exploration of anatomical structures, the image being obtained by an X-ray beam), or radiotherapy using an X or g beam (accelerator or bomb cobalt). The exposure decreases if we move away from the source and disappears if it is removed or if an effective screen is interposed. Other examples include by the presence of radioactive substances on the skin (we also speak of external contamination), for example, following contact with an object containing such free substances; to eliminate exposure, the body surface should be cleaned appropriately.

Internal exposure can occur in different ways: by inhalation of radioactive substances in the air; by ingestion of contaminated products (e.g. food); by transcutaneous penetration of external contamination (e.g. in the event of an injury or sore); during medical examinations such as diagnosis, scintigraphy (in this case, the image of the organ is obtained thanks to the emission g of the injected radioelement) and metabolic radiotherapy with the injection of radioactive substances.

Internal exposure lasts as long as the radioactive substances remain in the body; it decreases over time depending on the radioactive decay of the incorporated radioelements and their natural elimination by excretion (see: https://www.asn.fr/espace-professionnels/agrements-controles-et-mesures/effets-des-rayonnements-ionisants, 2021).

Deterministic effects are to be distinguished from long-term effects called stochastic or probabilistic characterized by an increase in the risk of cancer and genetic mutations. These stochastic effects are present not only in the domain of low doses but also in the domain of high doses. Indeed, we consider that each exposure presents a risk of stochastic effect whatever the dose received. The probability of triggering cancer is 5% per sievert. These probabilities are small or even almost zero when exposure becomes very low (see: https://laradioactivite.com/questionsdoses/echelle_doses, 2024).

2.1.3. *Advantages of technetiated tracers over thallium-201*

For myocardial stress scintigraphy (see section 2.7), technetiated tracers (such as ^{99m}Tc) are generally administered at a rate of 3.7 $MBq \cdot kg^{-1}$ body weight, then 11 $MBq \cdot kg^{-1}$ in protocols where stress and rest acquisitions are performed on the same day, and 11 $MBq \cdot kg^{-1}$ during stress and at rest in two-day protocols. For thallium-201, classically recommended activities are 1.5 $MBq \cdot kg^{-1}$ during stress and 0.5 $MBq \cdot kg^{-1}$ at rest [VER 11]. It should be noted that the physical properties of ^{99m}Tc are more advantageous than those of ^{201}Tl, as ^{99m}Tc has a half-life of around 6 h, reducing the dose delivered to the patient, with injected activities ranging from 10 to 20 mci compared to 2–3 mci for thallium, and its 140 keV energy is suitable for scintillation cameras [HAN 16]. Recalling that 1 ci = 3.7×10^{10} Bq, activities of 10–20 mci and 2–3 mci correspond to activities of 370 MBq–740 MBq and 74 MBq–111 MBq respectively.

2.1.4. *Uses of metastable technetium-99 in brain, bone and kidney scintigraphy*

Cerebral scintigraphy is an examination performed using a radioactive product. It enables cerebral activity to be visualized, highlighting the areas of the brain that are more irrigated when they are solicited. This examination is particularly useful for pinpointing areas of hyperactivity in epilepsy. The examination only requires an injection or perfusion [DEL 19].

Cerebral perfusion scintigraphy using technetized radiopharmaceuticals is an examination that uses a γ-camera to study the three-dimensional distribution of a cerebral perfusion tracer. Two radiopharmaceuticals are used in cerebral perfusion scintigraphy [TOU 01, ALA 08, SOC 19a, SOC 19b, JAN 15, HAU 19]:

– ^{99m}Tc-HMPAO (^{99m}Tc-**h**exa**m**ethyl**p**ropylene-**a**mine-**o**xime);

– ^{99m}Tc-ECD (^{99m}Tc-**e**thylenediyl bis-L-**c**ysteine-**d**iethyl-ester).

Bone scintigraphy is also performed by intravenous administration of a radiotracer using a carrier molecule from the biphosphonate (BPs) family, labeled with metastable technetium-99 (^{99m}Tc). The carrier molecule belongs to the BPs family, characterized by a P-C-P chain.

Depending on the possible substitutions on the central carbon atom, several molecules are available for use in bone scintigraphy: MDP (Methylene Di Phosphonate), EHDP (Ethylene Hydroxy Di Phosphonate), HMDP (Hydroxy Methylene Di Phosphonate) and DPD (Dicarboxy Propane Diphosphonate) [AGO 11].

Renal scintigraphy is a test used to visualize renal activity and detect any malformations of the kidney or elimination pathways (urethra, bladder). It is performed by injecting a radioactive product into the venous network. The examination will show the product collecting in the kidney and then being eliminated through the natural channels. It thus reveals obstacles to elimination, reflux, malformations or scarring and assesses kidney function [DEL 19]. The radiopharmaceuticals used in this scintigraphy are dimercapto succinic acid (DMSA) and *mercapto acetyl triglycine* (MAG_3) labeled with metastable technetium-99 (^{99m}Tc-DMSA and ^{99m}Tc-MAG_3).

Digestive scintigraphy, also known as *gastric emptying scintigraphy*, enables us to follow the passage of a meal through the stomach and to study stomach contractions. This examination is used in the event of digestive disorders, when a poor emptying of the stomach (called "gastroparesis") is suspected, for example, during diabetes. It can also be used to explore gastro-esophageal reflux or to search for gastroparesis, particularly in diabetic patients [NAL 14].

2.2. Metastable technetium-99 radiotracers

2.2.1. *Metastable technetium-99 production line*

Metastable technetium-99 is a nuclear isomer of technetium-99 produced by the ^{99}Mo-^{99m}Tc generator. ^{99}Tc is obtained directly (13%) from ^{99}Mo and indirectly (87%) via ^{99m}Tc [ARD 06, AUS 11, COU 18]. Technetium-99 is radioactive. It decays in the β^- mode to give stable ruthenium (Ru) 99. The equations for the nuclear reactions taking place in the ^{99}Mo-^{99m}Tc generator are described below.

By neutron bombardment, molybdenum-98 is transformed into molybdenum-99 with a half-life of 65.94 h (≈ 66 h), which, by β^- decay (82%) leads to metastable technetium-99 according to the respective equations:

$$ {}_0^1n + {}_{42}^{98}Mo \rightarrow \gamma + {}_{42}^{99}Mo \quad [2.3] $$

$$ {}_{42}^{99}Mo \rightarrow {}^{99m}_{43}Tc + {}_{-1}^{0}e \quad [2.4] $$

By *isomeric transition*, metastable technetium-99 is de-excited by emission of a photon γ of energy 140 keV to give technetium-99, according to the equation:

$$ {}^{99m}_{43}Tc \rightarrow {}_{43}^{99}Tc + \gamma \quad [2.5] $$

Figure 2.1 shows the production line of technetium-99m.

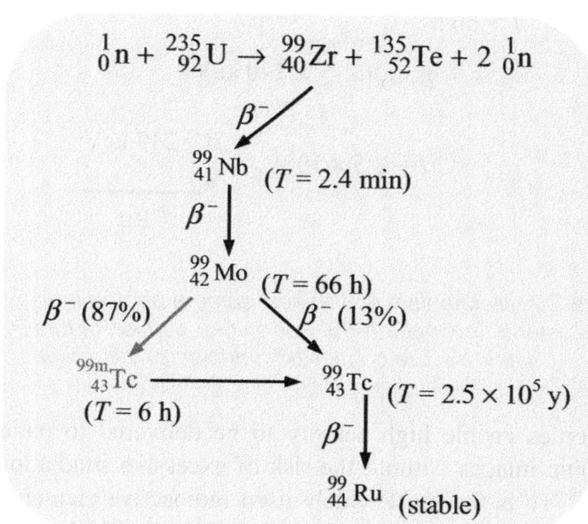

Figure 2.1. *Production line of technetium-99m via the ^{99}Mo-^{99m}Tc generator. For a color version of this figure, see www.iste.co.uk/sakho/nuclearphysics3.zip*

As shown in Figure 2.1, the half-life (6 h) of technetium-99m is relatively too short for storage [AUS 11]. For medical applications, it must be prepared on demand using the ^{99}Mo-^{99m}Tc generator, where ^{99m}Tc is extracted by elution with a sodium chloride solution at a concentration of 0.154 M (mol · L^{-1}) containing the species TcO_4^-, Na^+ at a concentration of less than 10^{-6} M. It should be noted that since the half-life of ^{99}Mo is equal to 66 h, the ^{99}Mo-^{99m}Tc generator can be used for 8–10 days [THO 18].

NOTE.–

In nuclear reactors, the fission of uranium-235 can also lead directly to the production of molybdenum-99, representing approximately 6% of fission products, according to the equation [AUS 11]:

$$^{235}_{92}U + {}^{1}_{0}n \rightarrow {}^{99}_{42}Mo + {}^{134}_{50}Sn + 3\,{}^{1}_{0}n + \gamma + Q\,(\text{heat}) \qquad [2.6]$$

2.2.2. *Preparation of metastable radiotechnetium-99*

From a medical point of view, technetium-99m is of great interest for scintigraphic analysis, as it is a pure γ-emitter of 140 keV (Figure 2.2), perfectly suited to γ-cameras with a 6 h half-life.

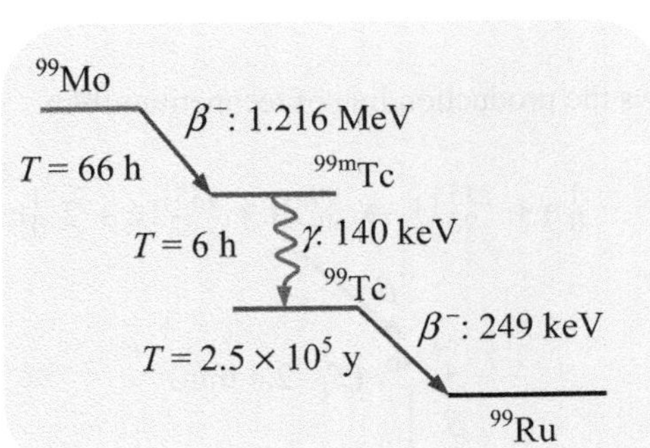

Figure 2.2. *Mechanism of 140 keV gamma photon emission from metastable technetium-99. For a color version of this figure, see www.iste.co.uk/sakho/nuclearphysics3.zip*

These properties enable high activity to be delivered to patients, resulting in good scintigraphic images without the risk of excessive irradiation [ARD 06]. For these reasons, ^{99m}Tc is the most widely used radioactive element in the world for single-photon emission computed tomography (SPECT) [DUC 15].

Technetium exists in various oxidation states, from (–I) to (+VII). In the presence of water, only the four states (+I), (+III), (+V) and (+VII) are stable; the other states only form in the presence of ligands other than water. In aqueous solution, at high potential and physiological pH (skin pH equal to 5.5), the stable form of Tc is that in the TcO_4^- (+VII), the only commercial form.

As shown in equation [2.5], ^{99}Mo decay via the β^- mode yields ^{99m}Tc. In practice, the ^{99}Mo-^{99m}Tc generator consists of an alumina column on which ^{99}Mo is fixed in the form of molybdate (molybdenum oxide MoO_3). Polymerized molybdate ($[^{99}Mo_7O_{24}]^{6-}$) strongly adsorbed on the ion exchange column decreases continuously, producing a mixture of ^{99m}Tc (87%) and ^{99}Tc (13%), as shown in Figure 2.2. This mixture is present as pertechnetate $Na^{99m}TcO_4$ and $Na^{99}TcO_4$ weakly bound to the column [ARD 06, POT 14, THO 18]. In these forms, the pertechnetate ions $^{99}TcO_4^-$ and $^{99m}TcO_4^-$ are easily eluted from the column by a sterile 0.9% aqueous sodium chloride solution, while the molybdate ion polymers are retained. The sterile system is enclosed in a lead shield to ensure radiation protection for medical staff.

The eluate is collected in a vacuum flask, allowing the sodium pertechnetate eluate to be aspirated. Due to the mixture of pertechnetate ions $^{99}TcO_4^-$ and $^{99m}TcO_4^-$, the eluate obtained does not consist of pure ^{99m}Tc. ^{99m}Tc is a 140 keV γ-emitter that can be detected by measuring instruments. ^{99}Tc can be considered stable since its very long radioactive half-life (2.5×10^5 y) compared with the half-life of ^{99m}Tc (6 h), makes it totally undetectable. Therefore, the main impurity in the sodium pertechnetate eluate is ^{99}Tc, which is of no diagnostic interest. The ^{99m}Tc production chain is shown in Figure 2.3.

Nuclear reactor
150 h to 200 h irradiation

99Mo extraction

^{99m}Tc generator production

5 h to 10 h for target release
4 h charging time
4 h to 15 h transport

12 h extraction of ^{99}Mo
12 h of transport to producers of ^{99m}Tc generators

12 h to 24 h for ^{99m}Tc generator production and transport to hospitals

Figure 2.3. *^{99m}Tc preparation line [COU 18]. For a color version of this figure, see www.iste.co.uk/sakho/nuclearphysics3.zip*

NOTE.–

In practice, generator elution can be fractional (elution carried out milliliter by milliliter) or global. In the former case, a volume activity (activity of the solution in relation to its volume) can be defined for each milliliter. In the latter case, the volume activity of the eluate represents an average value.

At the end of the production chain, the generator's elution yield is defined as the ratio between the ^{99m}Tc activity actually eluted and that actually present in the column. This ratio is an excellent indicator of generator quality. In practice, a margin of 10% is accepted on the elution yield, but a ratio of less than 100% may indicate a problem within the generator. This problem is often encountered when a generator has not been eluted for more than 48 h. In this case, ^{99m}Tc is not present in its usual oxidation state (+VII), but in a lower oxidation state [THO 18].

As explained above, the ^{99}Mo-^{99m}Tc generator consists of an alumina column to which ^{99}Mo is bound in the form of molybdenum oxide MoO_3. Radioactive decay within the generator column allows ^{99}Mo to be present in the pertechnetate eluate. Theoretically, ^{99}Mo should not contaminate the eluate, since it is not dissociated from the column during elution. However, from a practical point of view, alumina may be released during elution and contaminate the eluate. It is therefore imperative to determine the chemical purity of the eluate. This involves checking for the presence of alumina in the eluate. In practice, the mass concentration of alumina must be less than $5 \mu g \cdot mL^{-1}$ to be accepted. To achieve this, the activity of the eluate is measured unprotected and then with a six-millimeter-thick leaded vial guard. As lead is capable of completely blocking the 140 keV γ-radiation emitted by ^{99m}Tc (but not that of ^{99}Mo), the activity measured with the leaded vial shield will correspond to that of ^{99}Mo. It should be less than 1% of the activity measured without protection [HER 18, THO 18].

2.2.3. *Effective half-life, biological half-life, physical half-life*

By definition:

– The *effective half-life* is the time taken for the quantity of a specific radionuclide in the body to decrease by half its initial value, both through elimination by biological processes and through radioactive decay. This variable is widely used in radiotherapy.

– The *biological half-life* corresponds to the time after which half of an ingested or inhaled quantity (such as a drug or radioisotope) is eliminated from the body, or has lost its biological activity, by purely biological processes (metabolization, excretion).

– The *physical half-life* corresponds to the radioactive half-life. It is the time after which half of the initial nuclei of a radioactive element have decayed.

To express the relationship between these three half-lives, let us refer to T_e as the half-life or effective period, T_p as the half-life or physical period of the radioelement (radioactive period) and T_b as the half-life or biological period of the labeled compound. By definition [SAL 20]:

$$\frac{1}{T_e} = \frac{1}{T_p} + \frac{1}{T_b} \qquad [2.7]$$

The effective period can be deduced from this:

$$T_e = \frac{T_p T_b}{T_p + T_b} \qquad [2.8]$$

Formula [2.8] shows that the effective period is always smaller than the physical and biological ones.

APPLICATION 2.1.–

The physical half-life of ^{99m}Tc is 6.03 h and the biological half-life of exametazime-(^{99m}Tc) administered by direct IV route is approximately 48 h. What is the effective half-life of exametazime-(^{99m}Tc) in the indication of *brain scintigraphy*.

ANSWER.–

We use [2.8]. This gives:

$T_e = 6.03 \times 48/(6.03 + 48) = 5.36\ h \approx 5.4\ h$.

2.3. Technetium-labeled radiopharmaceuticals used in myocardial scintigraphy

2.3.1. *^{99}Mo-^{99m}Tc technetized radiopharmaceuticals*

Myocardial scintigraphy is an examination indicated in cardiology when coronary disease is suspected. It consists of studying the functioning of the heart, in particular its vascularization. Myocardial scintigraphy is particularly useful in patients with risk factors for coronary disease (diabetes, smoking, hypertension, high cholesterol) [DEL 19]. In the case of an organ such as the heart, two radiotracers are mainly used: thallium-201 (^{201}Tl) and metastable technetium-99 (^{99m}Tc), both of which bind to the myocardium [VER 11, HAN 16].

As we have already seen, the stable form of commercial technetium 99m coming out of the ^{99}Mo-^{99m}Tc generator is sodium perchlorate TcO_4^-, Na^+, in which ^{99m}Tc is in the +VII oxidation state. As this form is chemically inert, the perchlorate must first be reduced in order to obtain +I, +V and +IV oxidation states of ^{99m}Tc compatible with its complexing properties (the +V oxidation state is the oxidation state of Tc in the perchlorate ion TcO_4^-). This is achieved using stannous chloride dihydrate ($SnCl_2$, $2H_2O$), the most common reducing agent.

In practice, two technetized products are available. ^{99m}Tc sestamibi (Cardiolite®) and ^{99m}Tc-labeled tetrofosmin (Myoview®) [MEN 06, POT 14, AGE 15, CEY 19, THO 18]. The family of complexes with isonitrile ligands includes quantitatively obtained and highly stable complexes. Isonitriles are isoelectronic carbonyl groups and therefore generate very strong bonds with low-oxidation metals such as technetium. All of these complexes have an octahedral structure. The most common member of this family is ^{99m}Tc-Sestamibi [MAR 12a].

2.3.2. *Principle of ^{99m}Tc-MIBI scintigraphy and dosimetry*

Once prepared with an injectable solution of sodium pertechnetate, the ^{99m}Tc-sestamibi complex with the structure $[(^{99m}Tc)\,(MIBI)_6]^+$ (where MIBI = 2-methoxy-isobutyl-isonitrile) is formed [POT 14, MIN 17]. ^{99m}Tc-MIBI is a lipophilic cation (a substance with an affinity for apolar solvents such as fats) which enters the cell by passive diffusion and concentrates in the mitochondria. Its structure is shown in Figure 2.4.

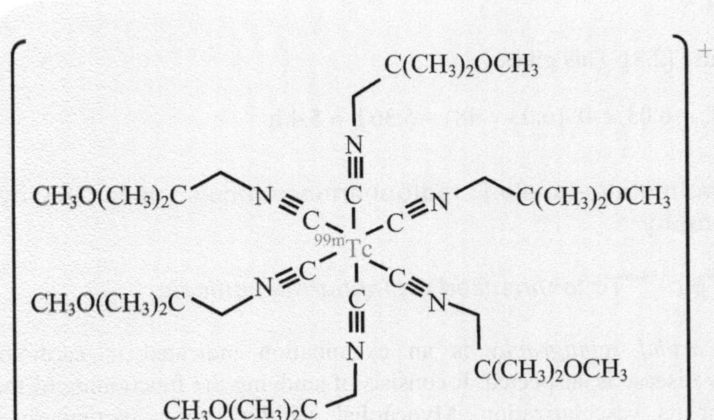

Figure 2.4. *Structure of ^{99m}Tc sestamibi or ^{99m}Tc-MIBI (2-Methoxy-IsoButyl-Isonitrile). For a color version of this figure, see www.iste.co.uk/sakho/nuclearphysics3.zip*

With regard to its pharmacokinetic properties, ^{99m}Tc-MIBI is rapidly distributed in tissues: 5 min after injection, around 8% of the injected activity persists in the circulation. Physiological distribution of ^{99m}Tc-MIBI includes myocardium, salivary glands, thyroid, myocardium, liver, gall bladder, colon, small intestine, kidney, bladder, choroid plexus, skeletal muscle and occasionally nipples. Low homogeneous uptake in breasts or armpits is normal [MIN 17]. After preparation with sodium pertechnetate (^{99m}Tc) solution, the dosage of ^{99m}Tc-MIBI solution depends on the patient's age, the type of organ to be diagnosed and the characteristics of the γ-camera used (conventional γ-camera or new generation CZT γ-camera, see Table 2.2).

For adults with an average body mass of 70 kg, the recommended activities for intravenous injection are as follows [AGE 15]:

– For the diagnosis of coronary perfusion defects and myocardial infarction: 400–900 MBq; for the diagnosis of myocardial infarction, an injection at rest is usually sufficient.

– For the diagnosis of myocardial ischemia, two injections (at the end of the stimulation test and at rest) are required to differentiate between transient and persistent myocardial hypofixation. Injections are made in doses of: 600–900 MBq for a two-day examination protocol and 400–500 MBq for a one-day examination protocol for the first injection and triple for the second injection, i.e. 1,200–1,500 MBq. In total, the activity administered must not exceed 2,000 MBq for a one-day examination protocol and 1,800 MBq for a two-day examination protocol. If the examination takes place over one day, the two injections must be performed at least 2 h apart, but the order is irrelevant. After the injection during the stimulation test, the patient is encouraged to continue physical activity for another minute (if possible).

^{99m}Tc-MIBI injected via a strict IV route is rapidly distributed in the tissues. Only approximately 8% of the injected activity is still present in the blood compartment 5 min after injection. The physiological distribution of ^{99m}Tc-MIBI includes several organs such as the myocardium, salivary glands, thyroid, liver, gallbladder, large intestine, small intestine, kidneys and bladder. Table 2.4 [TOU 01] shows the dosimetry used for ^{99m}Tc-MIBI scintigraphy.

Prior to the ^{99m}Tc-MIBI scintigraphy examination, the patient must be thoroughly prepared. To this end, they must be well hydrated before the start of the examination and asked to urinate as frequently as possible during the first few hours after the examination, in order to reduce radiation exposure. In the particular case of cardiac imaging, the patient should, if possible, have been fasting for at least 4 h at the time of the examination. It is recommended that they have a light snack of fatty food or drink one or two glasses of milk after each injection and before image acquisition. This precaution increases hepatobiliary clearance of ^{99m}Tc-MIBI, thus reducing hepatic activity during image acquisition. Acquisition is started approximately 30–60 min after injection, to allow hepatobiliary clearance of the product. Images can be acquired up to 6 h after injection, as no significant variation or redistribution of tracer concentration in the myocardium has been demonstrated [AGE 15].

^{99m}Tc	2-Methoxy-isobutyl-isonitrile (MIBI)				
	Intravenous injection				
	Absorbed dose per unit of activity administered (μGy/MBq)				
	At rest				
Organs	**Adult**	**15 years**	**10 years**	**5 years**	**1 year**
Gall bladder	39	45	58	100	320
Kidneys	36	43	59	85	150
Colon wall	24	31	50	79	149
Myocardium	6.3	8.2	12	18	30
Thyroid	5.3	7.9	12	24	45
Ovaries	9.1	12	18	25	45
Testes	3.8	5.0	7.5	11	21
Uterus	7.8	10	15	22	38
Effective dose (μSv/MBq)	**8.5**	**11**	**17**	**28**	**49**
	After effort				
	Absorbed dose per unit of activity administered (μGy/MBq)				
Organs	**Adult**	**15 years**	**10 years**	**5 years**	**1 year**
Gall bladder	33	38	49	86	260
Kidneys	26	32	42	63	110
Colon wall	19	25	41	64	117
Myocardium	7.2	9.4	10	21	35
Ovaries	8.1	11	15	23	40
Testes	3.7	4.8	7.1	11	20
Uterus	7.2	9.3	14	20	35
Effective dose (μSv/MBq)	**7.4**	**9.9**	**15**	**22**	**40**

Table 2.4. *Absorbed doses in ^{99m}Tc-sestamibi scintigraphy (MIBI). For a color version of this table, see www.iste.co.uk/sakho/nuclearphysics3.zip*

2.3.3. *Elimination of ^{99m}Tc-MIBI from the body*

^{99m}Tc-MIBI activity accumulated in the gall bladder is recovered in the intestine within 1 h of injection. Approximately 27% of the injected activity is renally eliminated within 24 h, and approximately 33% is eliminated in the feces within 48 h [AGE 15, MIN 17].

NOTE.–

An intravenous (IV) injection is the injection of a liquid into a vein, usually using a syringe and needle. In general, drugs are introduced into the body by various routes. They may be [LE 20]:

– swallowed (oral route);

– injected into a vein (intravenous, IV), into the muscle (intramuscular, IM), into the space around the spinal cord (intrathecal) or under the skin (subcutaneous, SC);

– placed under the tongue (sublingual route) or between the gums and the cheek (oral route);

– introduced into the rectum (rectal route) or vagina (vaginal route);

– instilled into the eye (ocular route) or ear (auricular route);

– sprayed into the nose and absorbed by the nasal membranes (nasal route);

– inspired into the lungs, usually through the mouth (inhalation), or through the mouth and nose (nebulization);

– applied to the skin (cutaneous route) for a local action (topical) or extended to the whole body (systemic);

– administered through the skin via a patch (transdermal) for a systemic effect.

Each injection route has its own specific objectives, advantages and disadvantages.

2.3.4. *^{99m}Tc-labeled tetrofosmin scintigraphy and dosimetry*

In addition to technetium (^{99m}Tc)-sestamibi, we also use ^{99m}Tc-labeled tetrofosmin, a cationic complex in the oxidation state + V that is also lipophilic and penetrates the cell by passive diffusion. Its structure is shown in Figure 2.5.

As with ^{99m}Tc-MIBI, ^{99m}Tc-labeled tetrofosmin can be used to diagnose or localize myocardial infarction or *myocardial ischemia*. For adults, two intravenous injections of ^{99m}Tc-labeled tetrofosmin are required. These two injections can be performed in any order, according to the two protocols of one day and two days. A first injection during exercise (or after pharmacological stimulation) and a second injection at rest.

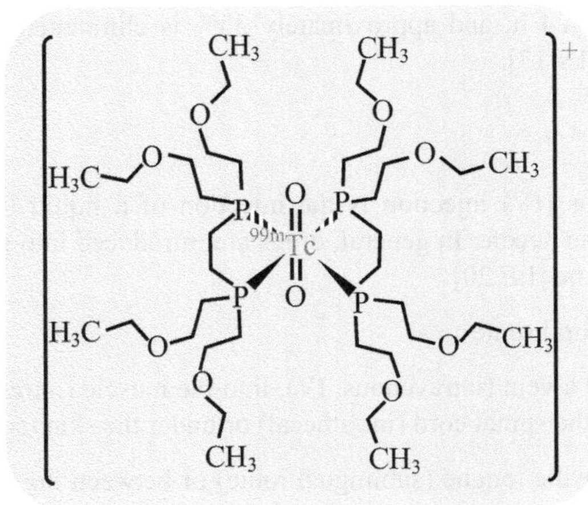

Figure 2.5. *Structure of ^{99m}Tc tetrofosmin. For a color version of this figure, see www.iste.co.uk/sakho/nuclearphysics3.zip*

When the two administrations are carried out on the same day, the activity of the first injection should be between 250 and 400 MBq, while the second will have an activity of between 600 and 800 MBq and should be administered at least 1 h later. If the two administrations are performed over two days, each injection should have an activity of between 400 and 600 MBq [THO 18].

^{99m}Tc	**Tetrofosmin (Myoview®)**				
	Intravenous injection				
	Absorbed dose per unit of activity administered (µGy/MBq)				
	At rest				
Organs	**Adult**	**15 years**	**10 years**	**5 years**	**1 year**
Gall bladder	36	40	53	93	310
Colon wall	24	30	49	78	143
Bladder wall	17	22	29	31	56
Myocardium	4.4	5.6	8.4	13	23
Ovaries	8.4	10	15	22	37
Testes	2.4	3.2	5.0	7.4	17
Uterus	7.2	9.0	14	20	32
Effective dose (µSv/MBq)	**6.8**	**8.7**	**13**	**20**	**36**

	After effort				
	Absorbed dose per unit of activity administered (μGy/MBq)				
Organs	**Adult**	**15 years**	**10 years**	**5 years**	**1 year**
Gall bladder	27	31	41	72	230
Colon wall	26	33	24	27	548
Bladder wall	18	23	36	57	107
Myocardium	4.8	6.1	9.0	14	24
Ovaries	7.6	9.5	13	19	31
Testes	2.9	3.9	5.1	7.7	13
Uterus	7.6	9.3	12	17	29
Effective dose (μSv/MBq)	**6.0**	**7.9**	**11**	**17**	**30**

Table 2.5. *Absorbed doses for ^{99m}Tc tetrofosmin scintigraphy (Myoview®). For a color version of this table, see www.iste.co.uk/sakho/nuclearphysics3.zip*

Unless otherwise indicated, the administration of ^{99m}Tc-labeled tetrofosmin is not recommended for young pediatric patients, as no studies have been carried out in this area. Table 2.5 [TOU 01] shows the dosimetry used for ^{99m}Tc tetrofosmin scintigraphy (Myoview®).

2.4. Radiopharmaceuticals labeled with metastable technetium-99 used in lung scintigraphy

2.4.1. *Technegas [^{99m}Tc] ventilation scintigraphy, dosimetry*

In addition to the metastable krypton-81 (^{81m}Kr) used as a radiotracer, aerosols labeled with metastable technetium-99 (^{99m}Tc) can also be used. Technetium 99m is produced by the ^{99}Mo-^{99m}Tc generator already discussed in section 2.2.1. An example of a solid aerosol (on the market since 1987) used to reveal pulmonary ventilation problems, and embolism in particular, is ^{99m}Tc-labeled Technegas®. Using a crucible (high-purity graphite) heated to between 2,500 and 2,550°C [LAC 11, MIN 21], under ultra-pure argon in the presence of sodium pertechnetate (^{99m}Tc), an aerosol of ultrafine carbon particles (size ranging from 5 to 200 nm) labeled with ^{99m}Tc is obtained, known as Technegas. It should be noted that the recommended activities of sodium pertechnetate (^{99m}Tc) to be deposited in the crucible are between 250 and 700 MBq in adults [MIN 21]. The dosimetry used for ^{99m}Tc Technegas® scintigraphy is presented in Table 2.6 [TOU 01].

	Technegas®				
^{99m}Tc	**Inhalation**				
	Absorbed dose per unit of activity administered				
	(µGy/MBq)				
Organs	**Adult**	**15 years**	**10 years**	**5 years**	**1 year**
Lungs	110	160	220	330	630
Myocardium	13	17	23	32	48
Esophagus	8.2	10	15	19	27
Thymus	8.2	10	15	19	27
Ovaries	0.41	0.55	1.1	2.0	4.2
Testes	0.06	0.09	0.20	0.33	1.1
Uterus	0.3	0.46	0.83	1.6	3.6
Effective dose (µSv/MBq)	**16**	**22**	**31**	**46**	**87**

Table 2.6. *Absorbed doses in Technegas® ventilation scintigraphy. For a color version of this table, see www.iste.co.uk/sakho/nuclearphysics3.zip*

^{99m}Tc Technegas® is administered by inhalation, no later than 10 min after preparation, as the carbon microparticles will tend to aggregate as soon as they are formed. Administration is via the "patient administration kit" containing a plastic tube. This tube is connected to the Technegas generator, which is fitted with a mouthpiece and filter. Medical staff administering Technegas® [^{99m}Tc] must wear disposable gloves. In addition, depending on the patient's condition (in the event of coughing and/or expectoration which may lead to the ejection of sputum or saliva), medical staff are asked to wear a cap and mask.

After each inhalation, the activity present in the lungs varies from patient to patient. It is recommended to monitor the lung count rate during Technegas inhalation, using a γ-camera equipped with a standard collimator (low energy, low/medium resolution), until a lung count rate of between 1.5 and 2 kcp/s (or often kcps: kilo counts per second) is reached. Inhalation should then be stopped. In adults, this corresponds to approximately 40 MBq of inhaled Technegas. The activity administered to children is a fraction of the activity recommended for adults [MIN 21].

Depending on the patient's choice, it is recommended that are taught to breathe through the mouthpiece using one of the following modes of administration:

1) deep, slow inhalation from functional residual capacity (end of calm exhalation), followed by a 5 s pause in inspiration (recommended method);

2) breathing deeply without pausing;

3) fast, deep breaths from functional residual capacity, followed by a pause of about 5 s on inspiration.

Patients with breathing difficulties may remove the mouthpiece between inhalations of Technegas. As the first inhalation of Technegas is oxygen-free, pre-oxygenation is recommended prior to Technegas administration, particularly in patients with respiratory insufficiency. To achieve uniform distribution, it is recommended that administration be carried out with the patient in the supine position [BOM 12, MIN 21].

As regards clearance, after inhalation, Technegas is adsorbed to the walls of the pulmonary alveoli and remains in the lung. There is no intravascular clearance, and radioactivity disappears by physical decay of Technetium-99m. Some of the carbon microparticles (from graphite) may be retained in the upper and central airways and are more significant in patients with airway obstruction or when inspiratory flow is insufficient (elderly people). These particles are eliminated via the digestive tract without being absorbed.

2.4.2. [^{99m}Tc]-DTPA ventilation scintigraphy and dosimetry

Technetium-99m-labeled diethylenetriamine pentaacetic acid (DTPA), ^{99m}Tc-DTPA, is the most widely used liquid radiopharmaceutical for lung ventilation scintigraphy. ^{99m}Tc-DTPA is administered using a nebulizer (aerosol generator), such as the Venticis® II pneumatic system (see Chapter 3, section 3.5 for further details). After the ^{99m}Tc-DTPA has been introduced into a nebulizer chamber, the Venticis® II system uses compressed air or oxygen to generate an aerosol with a *Median Mass Aerodynamic Diameter* (MMAD) of around 0.9 μm. However, it is important to note that the MMAD does not remain constant during product administration. In fact, due to the hydrophilic nature of liquid ^{99m}Tc-DTPA particles, they tend to aggregate on contact with moisture in the airways, thus increasing their size.

From a dosimetric point of view, an activity of 500–1,000 MBq generated in the nebulizer results in an activity of 50–100 MBq in the lungs. An extract of the radiation doses delivered to humans by the administration of technetium pentetate (^{99m}Tc) aerosol is shown in Table 2.7 [TOU 01].

After inhalation during ventilation lung scintigraphy, [^{99m}Tc]-DTPA rapidly diffuses through the pulmonary alveoli, due to its solubility in water and its small size of less than 500 daltons. It then enters the blood compartment and is eliminated by the kidneys, with an elimination half-life of less than 70 min. Note that: 1 dalton = 1/12 of the mass of a carbon-12 atom.

	TechneScan DTPA				
^{99m}**Tc**	Inhalation in aerosol form				
	Absorbed dose per unit of activity administered (μGy/MBq)				
Organs	**Adult**	**15 years**	**10 years**	**5 years**	**1 year**
Lungs	17	26	36	54	100
Breasts	1.9	1.9	3.3	4.8	7.8
Kidneys	4.1	5.1	7.2	11	19
Thyroid	0.99	1.7	2.7	4.4	7.8
Ovaries	3.3	4.1	6.1	8.9	15
Testes	2.1	3.1	5.2	7.9	15
Uterus	5.9	7.2	11	16	27
Effective dose (μSv/MBq)	6.1	8.4	12	18	35

Table 2.7. *Doses absorbed in ventilation scintigraphy using Technetium-99m [^{99m}Tc]-DTPA-labeled diethylenetriamine pentaacetic acid solution. For a color version of this table, see www.iste.co.uk/sakho/nuclearphysics3.zip*

The principle of lung scintigraphy is described in detail in Chapter 3.

2.4.3. ^{99m}Tc-MAA lung perfusion scintigraphy and dosimetry

As explained above, lung scintigraphy involves two phases: a ventilation phase, followed by a perfusion phase. For the perfusion phase, the radio pharmaceutical drug (RPD) used is ^{99m}Tc-labeled human macro-albumin aggregates (MAA). The RPD is prepared from a kit containing MAA and a solution of sodium pertechnetate ($^{99m}TcO_4^-$ + Na^+). MAA diameters range from 10 to 100 μm [THO 18, BLA 18].

ICRP (International Commission on Radiological Protection) publication no. 80 states that the effective dose of Technegas® is 0.015 mSv/MBq to the whole body and 0.011 mSv/MBq to the lungs. This is consistent with the data given in the

AMM: the dose delivered for an inhaled activity of 40 MBq is 0.6 mSv (for a 70 kg adult) [BOM 12]. Table 2.8 [TOU 01] shows the dosimetry applied in perfusion scintigraphy using ^{99m}Tc-MAA as the radiopharmaceutical for various organs.

^{99m}Tc	Aerosols with slow pulmonary clearance (Albumin)				
	Absorbed dose per unit of activity administered (µGy/MBq)				
Organs	**Adult**	**15 years**	**10 years**	**5 years**	**1 year**
Lungs	93	140	190	290	560
Bladder wall	8.2	16	24	36	65
Breasts	6.4	6.4	12	17	24
Ovaries	1.0	6.4	12	17	24
Testes	0.58	0.84	1.5	2.3	4.6
Uterus	1.7	2.1	3.4	5.2	9.5
Effective dose (µSv/MBq)	**14**	**26**	**28**	**43**	**81**

Table 2.8. *Absorbed doses in ^{99m}Tc-MAA perfusion scintigraphy. For a color version of this table, see www.iste.co.uk/sakho/nuclearphysics3.zip*

2.5. Radiopharmaceuticals used in brain scintigraphy

2.5.1. *^{99m}Tc-HMPAO and ^{99m}Tc-ECD radiopharmaceutical structures and dosimetry*

Cerebral perfusion scintigraphy uses a γ-camera to study the three-dimensional distribution of the radiopharmaceutical used in the brain. Two technetized radiopharmaceuticals are used [TOU 01, ALA 08, SOC 19a, SOC 19b, JAN 15, HAU 19, DEC 19]:

– ^{99m}Tc-HMPAO (^{99m}Tc-**h**exa**m**ethyl**p**ropylene-**a**mine-**o**xime);

– ^{99m}Tc-ECD (^{99m}Tc-**e**thylenediyl bis-L-**c**ysteine-**d**iethyl-ester).

Figure 2.6 shows the structures of ^{99m}Tc-HMPAO and ^{99m}Tc-ECD radiopharmaceuticals.

Comparative studies with other reference tracers, either of cerebral blood flow (with xenon 133) or of glucose consumption (with ^{18}F-fluoro-deoxy-glucose), suggest that HMPAO is a "true" perfusion tracer, whereas ECD, due to its enzyme-mediated binding mechanism, is more of a metabolic indicator. Indeed, HMPAO is capable of identifying luxury perfusions in the subacute phase of an ischemic stroke, unlike ECD [HAB 13].

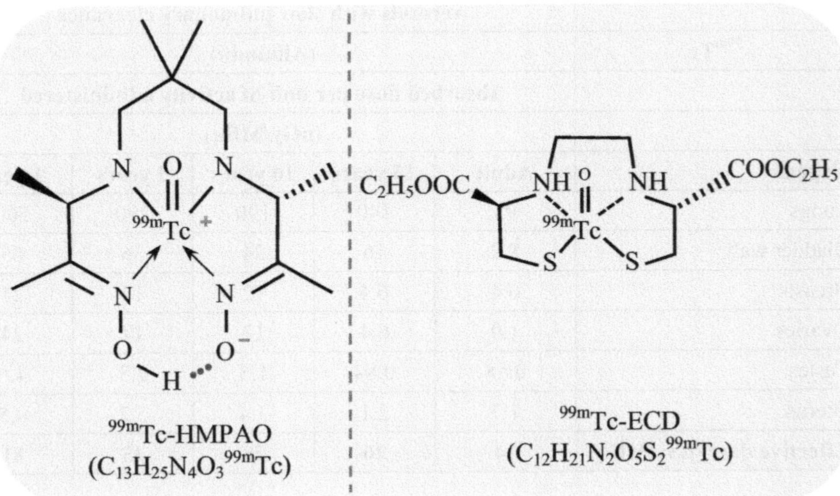

Figure 2.6. *Structures of ^{99m}Tc-HMPAO (^{99m}Tc-hexamethylpropylene-amine-oxime) and ^{99m}Tc-ECD (^{99m}Tc-ethylenediyl bis-L-cysteine-diethyl-ester). For a color version of this figure, see www.iste.co.uk/sakho/nuclearphysics3.zip*

For ^{99m}Tc-HMPAO, and for a maximum recommended injected activity of 1,110 MBq and a 70 kg adult, the effective dose is 10.3 mSv. For an injected activity of 740 MBq and a 70 kg adult, the absorbed dose to the target organ, the brain, is 5.0 mGy, and the absorbed dose to the critical organ, the kidneys, is 25.2 mGy. The effective dose is 6.88 mSv. In vitro stabilization with cobalt does not significantly alter the biodistribution and therefore the dosimetry of exametazime-(^{99m}Tc) [AGE 21]. Radiation doses absorbed by patients are presented in Table 2.9, according to the ICRP 80 publication [TOU 01].

Both radiopharmaceuticals are bound in the brain on first pass. Binding reflects perfusion and, depending on the tracer, the state of redox metabolism (HMPAO) or esterase enzyme activity (ECD) [SOC 19a].

	Exametazime (HMPAO)				
^{99m}Tc	**Intravenous injection**				
	Absorbed dose per unit of activity administered (μGy/MBq)				
Organs	**Adult**	**15 years**	**10 years**	**5 years**	**1 year**
Kidneys	34	41	57	81	140
Thyroid	26	42	63	140	260
Colon wall	17	22	35	55	101
Brain	6.8	11	16	21	37
Ovaries	6.6	8.3	12	17	27
Testes	2.4	3.0	4.4	6.1	11
Uterus	6.6	8.1	12	15	25
Effective dose (μSv/MBq)	**9.3**	**12**	**17**	**27**	**49**
	Bicisate (ECD)				
^{99m}Tc	**Intravenous injection**				
	Absorbed dose per unit of activity administered (μGy/MBq)				
Organs	**Adult**	**15 years**	**10 years**	**5 years**	**1 year**
Bladder wall	49	61	74	72	130
Gall bladder	27	31	40	70	230
Colon wall	19	24	39	60	115
Brain	4.9	8.0	14	19	31
Ovaries	7.5	9.4	13	17	27
Testes	2.7	3.6	5.2	6.1	11
Uterus	8.9	11	15	17	28
Effective dose (μSv/MBq)	**7.2**	**9.1**	**13**	**17**	**34**

Table 2.9. *Radiation doses absorbed by patients undergoing ^{99m}Tc-HMPAO or ^{99m}Tc-ECD brain scans. For a color version of this table, see www.iste.co.uk/sakho/nuclearphysics3.zip*

VOCABULARY CORNER.–

– *Luxury perfusion*: this describes blood flow in excess of local metabolic requirements to infarcted brain regions. After cerebral infarction, cerebral blood flow generally returns to normal or high levels compared with a normal brain [KNI 23].

– *Cerebrovascular accident (CVA)*: a neurological deficit of vascular origin. Strokes are the leading cause of death in women, the second leading cause of death in men, and the leading cause of acquired disability in adults. The average age of onset is 74, although middle-aged adults are increasingly affected. There are two main causes of strokes: ischemic strokes (80%) and hemorrhagic strokes (15%), which have a poorer prognosis. Transient ischemic attack (TIA) is a stroke caused by very transient arterial obstruction (< 1 h), with no sequelae of brain damage. It should be treated as an emergency. It may herald a more serious stroke. Its etiological assessment is often negative [HAU 22].

– *Ischemic stroke*: a stroke is typically due to obstruction of a cerebral artery, while a hemorrhagic stroke is due to rupture of a blood vessel, resulting in hemorrhage (see section 3.8.1 for more details on strokes).

– *Enzyme*: protein with catalytic properties.

– *Esterase*: a hydrolase enzyme that splits esters into an acid and an alcohol in a chemical reaction with water called hydrolysis.

– *Esterase activity*: a term used in chemistry to describe everything to do with esterase, a specific enzyme. Esterase plays a key role in biological processes, facilitating the hydrolysis of ester bonds. This hydrolysis transforms esters into the corresponding alcohols and acids.

– *Hydrolases*: a class of enzymes that catalyze the hydrolysis of molecules. Examples include esterases, which hydrolyze esters (R-CO-O-R').

– *Metabolism*: all of the chemical reactions that take place within each cell of a living being. In particular, it enables each cell to sustain itself, reproduce (divide), develop and respond to environmental stimuli (e.g. exchanges). A redox reaction is a chemical reaction in which electrons are transferred from an oxidizing agent to a reducing agent.

2.5.2. Characteristics of ^{99m}Tc-HMPAO and ^{99m}Tc-ECD radiopharmaceuticals

The comparative characteristics of ^{99m}Tc-HMPAO and ^{99m}Tc-ECD radiopharmaceuticals are shown in Table 2.10.

Comparative data	^{99m}Tc-HMPAO	^{99m}Tc-ECD
Cerebral extraction on first pass	About 80% extraction	Relatively moderate extraction at 57%,
Cerebral retention rate of injected activity	About 5%	About 6%
Blood circulation activity	Persistent activity of the order of 10–15%.	2% only 1 h after injection
Physiological elimination	10% by hepatic route and 35% by urinary route	Primarily by urinary route (73%)

Table 2.10. *Comparative characteristics of ^{99m}Tc-HMPAO and ^{99m}Tc-ECD*

In addition, in vivo stability of HMPAO is achieved rapidly, 2 min after having the radiotracer injection. HMPAO is also unstable in vitro and in vivo, as the lipophilic form degrades into a hydrophilic form. This transition to hydrophilicity is at the root of the intracerebral retention mechanism. In practical terms, this means injecting the tracer within 30 min of reconstitution and labeling. After that, the level of lipophilic molecules is no longer sufficient. A form stabilized by the addition of cobalt to the preparation allows 5 h to elapse between preparation and injection. As regards ECD, distribution is not absolutely stable, since there is considerable cerebral clearance, and it is recommended that images should not be taken more than 2 h post-injection. Tracer retention occurs by transformation of the lipophilic complex into a hydrophilic complex under the action of an esterase enzyme. ECD is stable 6 h after labeling, unlike HMPAO in its classic form [HAB 13]. In children, the administration of exametazime-(^{99m}Tc) stabilized with cobalt solution is not recommended. The safety and efficacy of cobalt solution has not been established in the pediatric population [AGE 21].

2.5.3. *Principle of the brain scan*

As seen, brain perfusion scintigraphy exploits the properties of two radiopharmaceuticals: ^{99m}Tc-HMPAO and ^{99m}Tc-ECD. Cerebral perfusion scintigraphy with HMPAO or ECD examines the brain's blood flow, highlighting the areas of the brain that are more irrigated when under stress. It can also be used to diagnose cognitive disorders, strokes and epilepsy. The examination only requires an injection or infusion [DEL 19, DEC 19].

In this section, we consider the special case of ^{99m}Tc-HMPAO, which is the most widely studied radiopharmaceutical in the literature consulted. This may be justified by the fact that HMPAO is considered a perfusion tracer, whereas ECD is more of a metabolic indicator.

NOTE.–

Epilepsy, from the Greek "epilambanein" (to seize, to attack by surprise), is defined by the recurrence of a symptom called an *epileptic seizure*. An epileptic seizure is the transient presence of signs and/or symptoms due to excessive, more or less synchronous, abnormal neuronal activity in the brain. It consists of three phases [LEP 18]:

– critical: during which the electrical discharge occurs, leading to the epileptic syndrome;

– post-critical: the period following the seizure, from a few minutes to a few hours;

– inter-critical: phase between two seizures, starting 24 h after the last seizure.

2.5.4. *Preparing and carrying out the cerebral perfusion scintigraphy examination*

The radiopharmaceutical ^{99m}Tc-HMPAO for injection is nontoxic. Furthermore, after injection, it does not cause any noticeable side effects such as malaise or drowsiness, and there is virtually no allergy to it.

Therefore, the cerebral perfusion scintigraphy examination requires no preparation. However, patients who are pregnant or have delayed menstruation should report this before the examination. If breast-feeding, this should be interrupted for 24 h [RIM 23]. The procedure is as follows [NAL 14, DEC 19, RIM 23]:

– The patient must remain motionless in the semi-darkness of a quiet room. Under these conditions, the number of cerebral stimuli will be reduced to a minimum, which is important for measuring resting cerebral perfusion correctly.

– Then, while the patient is lying on a bed, a catheter is inserted and a small quantity of ^{99m}Tc-HMPAO is injected into a vein in one arm. This product is used to visualize the distribution of cerebral blood flow. To ensure proper absorption of the product by the brain, the patient is asked to remain as relaxed as possible.

– 10–15 min after administration, the scan is performed. During the scan, the patient lies on the examination table with their head under the camera. A γ-camera takes pictures of their brain from different angles. Acquisition time is approximately 35 min. In a nutshell:

- total examination time: approximately 1 h;

- injections: 1 IV injection of ^{99m}Tc-HMPAO;

- passage under the γ-camera: 1 pass of 35 min;
- images are taken 10–15 min post-IV.

2.6. Production, use and dosimetry of radiothallium-201

2.6.1. *Radiothallium-201 production process*

Thallium-201 is a cyclotron product emitting X- and γ-radiation and decays to mercury-201 by electron capture [MEN 06, LAP 12, INS 13b]. The reactions involved in the production of ^{201}Tl are as follows.

Lead-201 is obtained by proton bombardment of thallium-203 in a cyclotron following the nuclear transmutation reaction ^{203}Tl (*p*, 3*n*) ^{201}Pb:

$$ {}_{1}^{1}H + {}_{81}^{203}Tl \rightarrow 3\ {}_{0}^{1}n + {}_{82}^{201}Pb \qquad [2.9]$$

Once formed, lead-201 decays in the mode to give thallium-201, according to the reaction:

$$ {}_{82}^{201}Pb \rightarrow {}_{81}^{201}Tl + {}_{1}^{0}e + \nu_e \qquad [2.10]$$

Particles emitted	Energy (in keV)	Emission percentage (%)
Electron	12.1	2.34
	15.8–17.4	16
	27.8–29.4	4.18
	52.8–55.7	9.77
	66.9	1.32
	84.1	15.8
	121–153	3.51
Photon X	11.6–13.9	18.7
	69.1–71.1	73.7
	80.1–82.8	18.6
Photon γ	135	2.57
	166–167	10.2

Table 2.11. *Main thallium-201 emissions*

Thallium-201, with a half-life of 3.04 days (72.96 h), has a mass activity of 7.88×10^{15} Bq · g^{-1}. It decays by electron capture to stable mercury-201, emitting X-rays, according to the equation:

$$^{201}_{81}Tl + x\,^{0}_{-1}e \rightarrow \,^{201}_{80}Hg + X + \gamma + \nu_e + y\,^{0}_{-1}e \qquad [2.11]$$

The characteristics of the electromagnetic radiation (X and γ) and electrons emitted by ^{201}Tl are presented in Table 2.11 [INS 13b].

2.6.2. Use of ^{201}Tl radiotracer and dosimetry

In practice, the radiopharmaceuticals most commonly used for myocardial scintigraphy are ^{201}Tl (73 h half-life) and ^{99m}Tc-MIBI (6 h half-life). Because of its very long half-life, ^{201}Tl has the advantage of being able to perform imaging over a period of several hours, exploiting X-ray radiation at 80 keV [MEN 06, LAP 12]. As an injectable liquid, 1 mL of solution contains 37 MBq of thallous chloride (^{201}Tl) at calibration date. Specific radioactivity is greater than or equal to 3.7 MBq/µg of thallium [AGE 20b].

^{201}Tl	**Thallium chloride [1]**				
	Intravenous injection				
	Absorbed dose per unit of activity administered (µGy/MBq)				
Organs	**Adult**	**15 years**	**10 years**	**5 years**	**1 year**
Kidneys	480	580	820	1,200	2,200
Bone surface	340	440	720	1,200	2,900
Colon wall	235	302	512	850	1,616
Thyroid	220	350	540	1,200	2,300
Myocardium	200	260	390	620	1,100
Ovaries	730	620	2,000	3,500	8,300
Testes	450	1,100	8,300	9,600	13,000
Uterus	51	63	100	150	270
Effective dose (µSv/MBq)	**231**	**319**	**1,265**	**1,724**	**2,940**

Table 2.12. *Absorbed doses from ^{201}Tl thallium chloride scintigraphy[1]. Absorbed dose estimates do not take into account the presence of radiocontaminants. In fact, given the radiochemical purity of radionuclides currently available, the influence of radiocontaminants on effective dose is minimal. For a color version of this table, see www.iste.co.uk/sakho/nuclearphysics3.zip*

Clinically, thallium-201 chloride is used vector-free for myocardial scans to assess coronary perfusion and cell viability. ^{201}Tl is injected intravenously using a thallium chloride solution; the patient must have fasted for 4 h prior to injection. In general, two injections are required, one at rest and the other during exercise (or after pharmacological stimulation). For the first injection, activity will be 110 MBq, while for the second injection, ^{201}Tl activity will be 37 MBq. Image acquisition can begin a few minutes after the second injection. Table 2.12 [TOU 01] shows the dosimetry used for thallium-201 scintigraphy in the form of a thallium chloride solution.

2.6.3. *Excretion of thallium-201 from the body*

Clearance (purification at constant volume) is essentially via the feces and urine. After entering the bloodstream, thallium-201 is instantly distributed to the various tissues. 3% of the activity transferred to the blood is retained in the kidneys over a period of 10 days, and the rest is uniformly distributed to the other organs over the same period [INS 13b]. The myocardial extraction fraction of thallium-201 is around 85% on first pass. Myocardial activity reaches a maximum of 4–5% of the injected activity and remains relatively constant for 20–25 min. In terms of clearance, thallium is mainly eliminated in the feces (80%), with the remainder in the urine (20%). Its effective half-life is 60 h, its biological half-life is around 10 d and its physical half-life is 73.1 h [AGE 20b].

APPLICATION 2.2.–

The biological half-life of thallium-201 is 10 days. Determine its effective half-life.

ANSWER.–

The physical half-life of thallium-201 is 73 h (see Table 2.1). According to [2.8], its effective half-life is equal to 55.97 h ≈ 56 h.

APPLICATION 2.3.–

On Monday at 10 a.m., a nuclear medicine department orders a vial of radiothallium-201 with an activity of 296 MBq. This vial is to be used for six myocardial scans on the following Thursday at 10 a.m. It is assumed that all six IVs are performed simultaneously. Each of the six patients receives an activity of 110 Mbq. Determine the residual activity of the vial after the six IVs.

ANSWER.–

Let us find the residual activity after the six simultaneous injections. The general law of decay is:

$$A\,(t) = A_0 e^{-\lambda t} = A_0 e^{-\ln 2t/T} \qquad [2.12]$$

The injections are performed at the initial time on Thursday at 10 a.m. A_0 is therefore the initial activity required to perform the six myocardial scans on Thursday. The activity ordered on Monday at 10 a.m., four days earlier, is $A\ (t = 4\text{d}) = 296$ MBq. Using [2.12], we obtain the initial activity A_0, i.e.:

$$A_0 = A\,(t = 4\text{j})\, e^{\ln 2t/T} \qquad [2.13]$$

The activity required for the six IVs is $A_{IV} = 660$ Mbq. The residual activity A_r after the six simultaneous injections is then:

$$A_r = A_0 - A_{IV} = A\,(t = 4\text{j})\, e^{\ln 2t/T} - A_{IV} \qquad [2.14]$$

ANSWER.–

$$A_r = 296\, e^{4\times 24\times \ln 2/73} - 660 = 736.4898 - 660 = 76.4898 \text{ MBq} \approx 76.5 \text{ MBq}.$$

Note that the initial activity $A_0 \approx 736.5$ MBq is only sufficient to perform a total of six scans at a rate of 110 MBq per patient.

APPLICATION 2.4.–

The scintigraphy test from the previous application is reconsidered. After the six injections, the vial is stored in the radioactive waste room for n physical periods of ^{201}Tl. Determine the theoretical quenching time of the vial.

ANSWER.–

Theoretically, the flask is extinguished if it contains no more thallium-201 nuclei. Let N_r be the number of residual nuclei. The number of nuclei likely to decay is:

$$N\,(t) = N_r e^{-\ln 2t/T} \qquad [2.15]$$

For $t = nT$, we obtain (remember that: $a^{\alpha} = e^{\alpha \ln a}$):

$$N\,(t = nT) = N_r e^{-n \ln 2} = 2^{n} N_r$$

Let (given that $A = \lambda N$):

$$n = \ln[N_r/N\,(t = nT)]/\ln 2 = \ln[A_r/A\,(t = nT)]/\ln 2 \qquad [2.16]$$

The source is extinguished if the vial contains no more radioactive nuclei. There is no such thing as half a nucleus. Let us assume that the source is extinguished when $N\,(t = nT) = 0.5$. The corresponding activity $A\,(t = nT) = 0.5\,\lambda$. According to [2.16], this gives:

$$n = \ln[A_r/0.5\lambda]/\ln 2 = \ln[TA_r/(0.5\times \ln 2)]/\ln 2 \qquad [2.17]$$

DIGITAL APPLICATION.–

$$n = \ln[73\times 3{,}600\times 76\,000/(0.5\times\ln 2)]/\ln 2 = 22.47 \approx 22$$

The source is theoretically extinguished at date $t = 22T = 1{,}606$ h $= 66.97$ d, i.e. approximately two months and one week after the vial is stored in the radioactive waste room. This two-month period is reasonable. However, the vial must be placed in an enclosure made of a material capable of absorbing all the ionizing radiation (electrons, X-ray photons and photons) emitted by thallium-201 (see Table 2.11). In the general case, a radioactive source is theoretically extinguished if its activity $A = \lambda N = 0.5\lambda$.

APPLICATION 2.5.–

The thallium-201 solution contained in the previous vial received on Thursday at 10 a.m. contains the isotope ^{202}Tl, an impurity whose activity is 0.5% times that of ^{201}Tl. Knowing that the solution can no longer be used when the activity ratio $A(^{202}Tl)/A(^{201}Tl)$ is greater than 2%, determine the time after which the thallium-201 solution can no longer be injected. The half-life of ^{202}Tl is 294 h.

ANSWER.–

Let $A(^{201}Tl) = A_1$ and $A(^{202}Tl) = A_2$.

$$\begin{cases} A_1(t) = A_{01}e^{-\ln 2t/T_1} \\ A_2(t) = A_{02}e^{-\ln 2t/T_2} \end{cases} \Rightarrow \frac{A_2(t)}{A_1(t)} = \frac{A_{02}}{A_{01}}\frac{e^{-\ln 2t/T_2}}{e^{-\ln 2t/T_1}} = \frac{A_{02}}{A_{01}}e^{\ln 2t(1/T_1 - 1/T_2)}.$$

This results in:

$$2\% = 0.5\% \times e^{\ln 2t(1/T_1 - 1/T_2)}$$

Therefore:

$$t = \frac{\ln 4}{\ln 2}\frac{T_1T_2}{T_2 - T_1} = 2\frac{T_1T_2}{T_2 - T_1} \qquad [2.18]$$

DIGITAL APPLICATION.–

$$t = 2\frac{73 \times 294}{294 - 73} = 194.23 \text{ h} \approx 8 \text{ d } 2 \text{ h } 6 \text{ min.}$$

In practice, the thallium solution is no longer usable eight days after it is received in the nuclear medicine department at 10 a.m. on Thursday.

2.7. Myocardial scintigraphy

2.7.1. *Definition and uses*

Myocardial scintigraphy is an examination indicated in cardiology when coronary disease is suspected. It consists of studying the functioning of the heart, in particular its vascularization. Myocardial scintigraphy is particularly useful in patients with risk factors for coronary disease (diabetes, smoking, hypertension, high cholesterol) [DEL 19].

Myocardial scintigraphy or *cardiac scintigraphy* is also known as *stress perfusion scintigraphy*. The examination consists of an exercise test on a bicycle or treadmill, or a pharmacological stress test sensitized by the injection of, for example, Dipyridamole (Persantine®) (see section 2.7.7). The test is carried out under the constant supervision of a cardiologist, whose role is to assess the indication and modality of the stress test. The cardiologist also monitors the patient to prevent and treat any complications (e.g. pain, shortness of breath, fatigue, malaise, etc.) that may arise during the test.

Myocardial scintigraphy is an examination prescribed by a cardiologist to a patient to assess myocardial perfusion, i.e. the quality of blood flow through the heart's coronary arteries (section 2.7.8). This examination also provides a number of fundamental insights into the functioning of the heart muscle, such as:

– its overall contractility, which provides information on the heart's pump function;

– its local contractility relative to the ventricular blood ejection fraction;

– the study of cellular metabolism, in particular to assess myocardial viability. For example, following a myocardial infarction, myocardial scintigraphy can be used to check whether there are still viable areas of the infarct that can still recover some contraction;

– the study of the repercussions on heart function of lesions such as narrowing of the coronary arteries (coronary artery bypass grafting, also known as aorto-coronary bypass grafting, can be applied: this is a cardiac surgery technique consisting of bypassing a narrowed or blocked coronary artery by implanting another vessel downstream from it);

– the search for defects in cardiac perfusion in cases of suspected angina pectoris, heart failure or coronary dilatation.

2.7.2. *Procedure*

Stress myocardial scintigraphy is performed in two phases: one during exercise and the other at rest. The procedure is summarized below [NAL 14]:

– patient preparation: no fasting required, but no coffee, tea, chocolate or bananas in the morning before the stress test (interaction with Persantine® or Rapiscan®);

– total exam time: 4 h;

– injection: one or two injections;

– passages under the γ-camera: one to two passes of 20 min each.

In addition, the specific modalities of the stress test are as follows [NAL 14]:

– IV injection of radiotracer at the acme (*end*) of the effort or stress;

– stress images taken a few minutes (15–30 min) after the test under the γ-camera. If the images are strictly normal, the examination is stopped after this series of images;

– new IV injection of radiotracer for resting images 3–4 h after exercise;

– resting images taken under the γ-camera 15–30 min after injection.

As we saw in Chapter 2, myocardial scintigraphy involves the use of radiotracers, the γ-radiation emitted by these radiotracers being captured by a γ-camera. The radiotracers used are thallium-201 (^{201}Tl) and metastable technetium-99 (^{99m}Tc), which binds to the myocardium.

2.7.3. *Ergometric bicycle and treadmill*

Ergometry is the science of measuring the work done by certain muscles or parts of the body (ergometric bicycle and treadmill). Exercise testing remains an essential part of general cardiology. It enables us to explore the body's capacity to adapt. This is so important since the physical capacity explored by the stress test is currently the best independent marker of risk factors and associated diseases for individual life expectancy [DOU 20].

The advantages of the ergometric bicycle are due to the fact that exploration on a bicycle brings greater stability to the ECG (and therefore greater ease of interpretation), easier measurement of blood pressure (BP) and greater safety (lower risk of falling). The ergometer also makes exercise easier for overweight subjects, even though the energy cost estimated by oxygen consumption (VO_2) is 5–6 mL/min/kg higher (the need to mobilize heavier legs) than for subjects of normal morphology.

The advantages of the treadmill are linked to the fact that in 80% of subjects, performance on the treadmill is 5–11% higher than on a seated bike. Maximum heart rate (HR max) is also often higher in American studies, but no difference is found in a few European studies (Switzerland) with identical HR max on bike and treadmill, despite a higher VO_2 peak (+7%) on treadmill. A difference of 5–10 beats between the two types of exercise test is therefore not automatic [DOU 20]. Figure 2.7 illustrates a cardiologist-assisted treadmill exercise test [RIM 21]. The case of an exercise test on a bicycle will be presented in Figure 2.7.

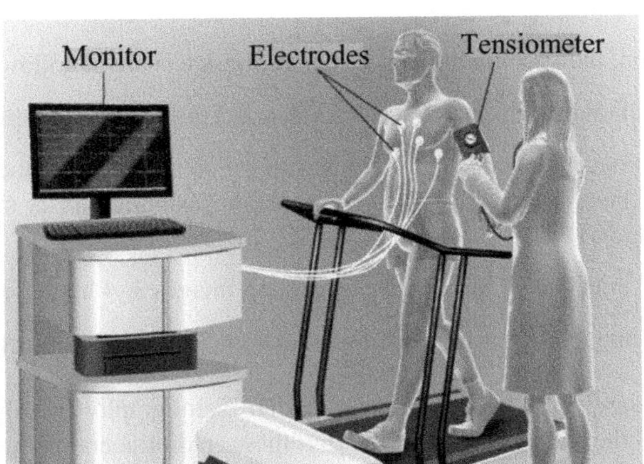

Figure 2.7. *Treadmill stress test. Image taken and adapted from the source: https://clinique-de-chatillon.ramsaysante.test-deffort-cardiaque. For a color version of this figure, see www.iste.co.uk/sakho/nuclearphysics3.zip*

Electrodes applied to the patient's chest (Figure 2.7) monitor the heart rate (at monitor level) under the responsibility and continuous supervision of a cardiologist. The cardiologist is responsible for providing information on the progress of the stress test.

2.7.4. *Performing the stress test*

Before the test is carried out, the patient must be informed that the *stress test* will be performed to the best of their ability and that they must be able to remain motionless during the *tomoscintigraphic acquisitions* (from 2 to 15 min, depending on the protocol used and the type of camera). It is preferable for the patient to have been fasting since the evening before if the protocol begins with a resting acquisition or to have had a light breakfast if it begins with a stress acquisition. Patients should be advised to avoid all stimulants (tea, coffee, chocolate, banana or cola) for 12 h prior to the test.

In addition, in view of the associated stress test, the patient is advised to bring suitable clothing (shorts, tracksuit, sports shoes). Women must report any pregnancy or late menstruation, as well as any breast-feeding. Asthma must also be reported. If the examination also includes a resting scintigraphy on the same day, this will be communicated by the cardiologist. If the patient is coming from far away, it is a good idea to bring something to eat and drink. Depending on the case, the attending cardiologist may ask the patient to stop taking certain medications 24–48 h before the examination (e.g. beta-blockers, which slow down the heart and reduce the sensitivity of the technique) [NAL 14, RIM 21].

The stress test is organized as follows [VER 11]:

1) The patient usually starts the effort at a load of 20 Watts (in the case of an ergometric cycle test), then goes through successive 20-Watt stages until the maximum achievable effort is reached.

2) The patient is followed by cardiac monitoring (blood pressure, continuous ECG, heart rate). A cardiologist is present throughout the examination, to ensure that it is carried out correctly and to determine the criteria for stopping the effort.

These criteria must be mentioned in the final report. This is followed by a recovery period lasting from one to several minutes. Physical effort, which is to be preferred, acts both on the functioning of the heart, by increasing its workload and oxygen consumption, and on arteriolar functioning, particularly coronary, by vasodilatation.

The three phases of the stress test are as follows.

First phase: stress test

Once the product has been set, the patient undergoes an exercise test lasting 15–30 min, consisting of pedaling on a bicycle (ergometric cycle) or walking on a treadmill at a relatively intense pace [VER 11, HAN 16]. The patient is asked to perform a maximal effort, defined by a heart rate equal to at least 85% of the theoretical maximum rate (220 – age). The radiotracer is injected at maximum effort.

Note that maximum heart rate (MHR) is the rate that a given person's human heart reaches under the highest demands. At rest, the heart rate is around 60–80 beats per minute. The maximum heart rate depends on age and is estimated using various formulas. Often, we use that of [FOX 71]:

$$\text{MHR} = 220 - \text{age “220 minus the age”}$$

Therefore, a 40-year-old man or woman will have an HRt (Theoretical Maximum Heart Rate) equal to 220 – 40 = 180. The MHR enables an athlete to know how far to push their physical effort. If an individual exceeds their estimated MHR, then we deduce that their maximum heart rate is higher than the estimate based on the formula applied. Generally speaking, there is no reason to stop an individual from exceeding their estimated MHR. If an individual is in good health (i.e. does not suffer from cardiovascular disease), then they can exceed their MHR without damaging their health.

In the general case, studies carried out to assess the optimal duration of exercise testing show that in only 12 healthy subjects, on treadmill and bicycle, suggest that exercise tests should have a duration of 10 min on average to obtain maximum performance. However, as this "optimal" duration is repeated in all recommendations, protocols must be individualized to meet this time criterion [DOU 20]. Figure 2.8 shows an example of an exercise test on a bicycle [HAN 16].

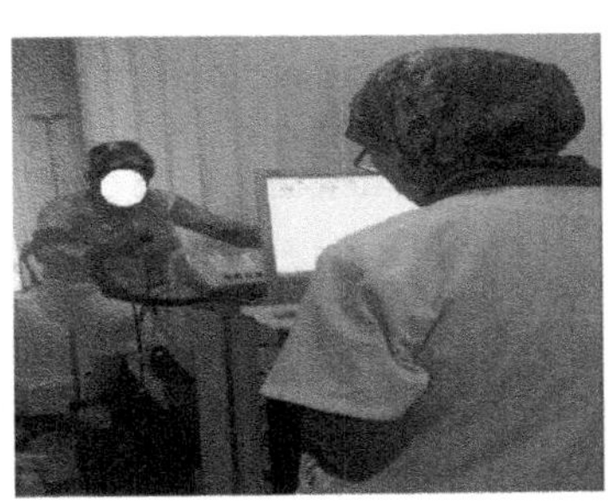

Figure 2.8. *Patient on bicycle (stress test). For a color version of this figure, see www.iste.co.uk/sakho/nuclearphysics3.zip*

A first series of scintigraphic images is recorded immediately after exercise, then during the recovery phase.

Second phase: acquisition of a series of stress images

Stress images are acquired starting 15 min after tracer injection. Acquisition takes around 10 min. While these images are being taken, the patient is usually lying on their stomach on an examination bed, under a γ-camera that requires total immobility (Figure 2.9) [HAN 16].

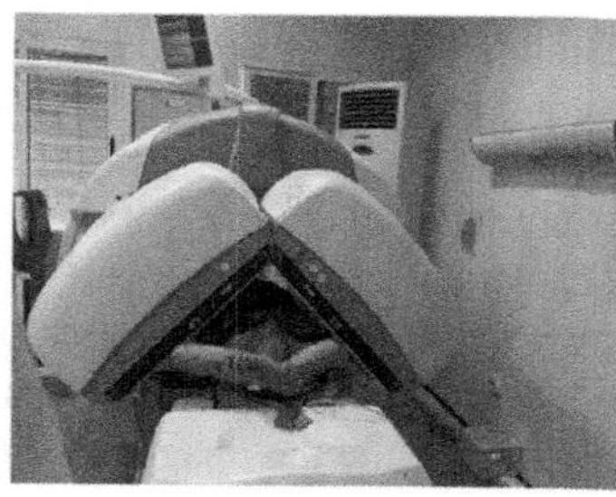

Figure 2.9. *Patient on cardiac scan examination bed. For a color version of this figure, see www.iste.co.uk/sakho/nuclearphysics3.zip*

Studying the distribution of the radiotracer in different parts of the heart provides information on the vitality of the tissue explored. This distribution appears as scintillating dots colored either red if the tissue is alive or blue-green if the tissue is dead.

2.7.5. *Arm exercise test and exercise intensity*

It should be noted that the exercise test does not have to be performed on a bicycle with the use of the legs, as shown in Figure 2.8. In fact, this test can be performed with the use of the arms for patients who have difficulty pedaling with their legs. However, in this case, performance is generally much lower (– 25% on average) than with the legs. This is because the muscle mass involved is lower, most subjects are often poorly trained with the arms, and performance is poorer. Interpretation of the results is, of course, difficult [DOU 20].

With regard to the intensity of exercise during the stress test, a warm-up period of 3 min on average, at low intensity is necessary in all cases (ranging, for example, from "no load" to already significant power levels of 80 or 100 watts for subjects trained on a bicycle). The increase in exercise intensity will be gradual, depending on the case, from 5 to 30 watts/min on a bicycle [DOU 20].

2.7.6. *Examination at rest*

A resting examination (without stress test) may also be necessary to rule out the presence of a narrowing that could have an impact on myocardial blood flow (if the stress scan is abnormal or doubtful). In addition, the resting test may be performed before the stress test (a very low dose will be injected), depending on the schedule and the presence of the cardiologists, and to avoid the need for the patient to return on another day if they are coming from far away.

During the resting examination, the patient lies comfortably, often on the stomach, under a CZT γ-camera, enabling the heart to be visualized in just a few minutes. Image acquisition begins 15–30 min after intravenous injection of the radiopharmaceutical ^{99m}Tc-MIB. During this initial phase, the patient is asked not to speak or move, however slightly, while the images are being taken.

2.7.7. *Myocardial scintigraphy coupled with pharmacological stimulation*

As we pointed out at the end of section 2.3.4, in order to make the examination as effective as possible, myocardial scintigraphy can be coupled with *pharmacological stimulation* using drugs [NAL 14, CEN 21a, RIM 21]. This examination is indicated in cases where the stress test would not be feasible (physical inability to pedal or run on a treadmill) or if the effort is not sufficient (patient not very active or not very motivated to move). In such cases, the test is made more sensitive by injection of Dipyridamole (Persantine®), for example.

Dipyridamole (Figure 2.10) is a drug that is injected into a vein in the arm and temporarily increases the diameter of the coronary arteries. In this way, it has the same effect on the heart as physical exertion. Injecting the drug causes stress to the patient's heart. The drug injected causes blood vessels to dilate. Minor side effects can include headaches, hot flushes and dizziness.

At the end of the stress test, the patient is admitted to the examination room for stress scan imaging. In this case, the patient will lie as comfortably as possible on a γ-camera table (Figure 2.10). This position is painless, and the patient is asked to remain silent and not to make the slightest movement while the images are being taken. This phase of transient immobility is a key criterion for image quality. As a general rule, images will be recorded during the recovery phase 15–30 min after the radiotracer injection and, finally, during rest 3–4 h later.

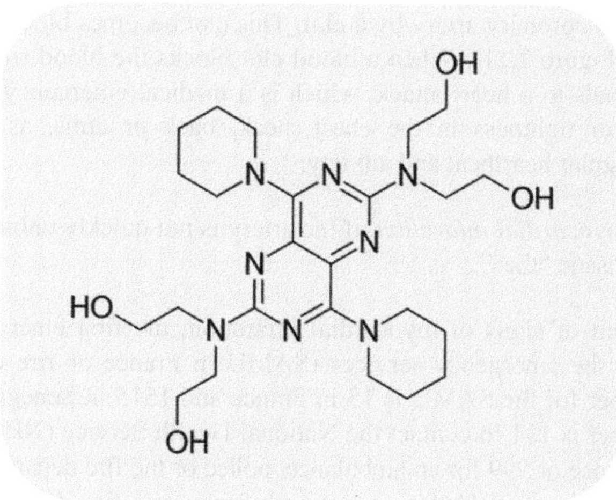

Figure 2.10. *Structure with empirical formula $C_{24}H_{40}N_8O_4$*

2.7.8. *General information on myocardial infarction*

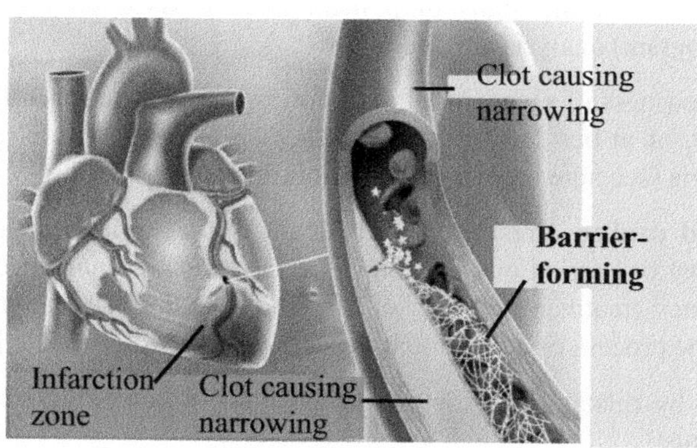

Figure 2.11. *Myocardial infarction due to obstruction of a coronary artery by a blood–barrier clot (source: [RAM 20]). For a color version of this figure, see www.iste.co.uk/sakho/nuclearphysics3.zip*

Generally speaking, infarction refers to myocardial infarction (the heart muscle). However, an infarction can affect any vascularized organ in the body as soon as the arterial circulation feeding the cells of that organ is interrupted. Myocardial infarction (or a heart attack) is defined as the necrosis of a greater or lesser part of the heart muscle. Myocardial death occurs when the heart muscle is no longer supplied with blood oxygen by the coronary arteries [FÉD 20]. This is caused by the obstruction of a coronary artery by a clot. This clot becomes blocked at a narrowing in the artery (Figure 2.11). When a blood clot blocks the blood supply to the heart, this usually leads to a heart attack, which is a medical emergency. Symptoms may include pain or tightness in the chest, neck, back or arms, as well as fatigue, dizziness, irregular heartbeat and anxiety.

During a *myocardial infarction*, if the artery is not quickly unblocked, part of the heart muscle tissue "dies".

In the event of signs of myocardial infarction, the first emergency measure is to alert (call) the emergency services (SAMU in France or fire department). The toll-free number for the SAMU is 15 in France and 1515 in Senegal. In the UK, the toll-free number is 111 to contact the National Health Service (NHS) for emergency medical assistance or 999 for an ambulance, police or the fire department. In the USA, the toll-free number is 911 for police, ambulance and fire departments. The rapid arrival of the emergency medical services (EMS) or fire department will enable the diagnosis to be made and the patient to be referred to a specialized center. The sequence of events relating to the assertion and severity of the reported infarction is as follows:

– emergency physicians and cardiologists have the prerogative to urgently confirm or rule out the signs of a heart attack, based on the results of ECG (electrocardiogram) analysis and blood tests;

– the seriousness of myocardial infarction is linked to the fact that it can rapidly lead to death, or at best, leave after-effects such as heart failure (see note below), angina pectoris (see note below) or arrhythmias (irregular rhythm);

– to avoid cardiac arrest, of which death is the logical outcome, emergency treatment must be carried out as soon as an infarction is confirmed within 90 min. This emergency treatment consists of "freeing" the coronary artery (Figure 2.11), which is in the process of becoming blocked, as quickly as possible [RAM 20]:

– at best, by dilatation of the artery, generally followed by stenting to prevent local recurrence,

– alternatively, by injecting a thrombolytic drug into a vein to dissolve the clot.

The next steps are to:

– take lifelong medication to limit the after-effects of a heart attack;

– avoid recurrence of thrombosis (clogging) with long-term aspirin therapy; in the case of a stent, another drug (antiplatelet agent) will also be given;

– combat risk factors: stop smoking, treat high cholesterol, fight excess weight, take part in physical activity.

Heart failure is a condition in which abnormal cardiac function is responsible for the myocardium's inability to deliver sufficient cardiac output to meet the body's energy requirements. This failure may reflect an anomaly in ventricular cardiac muscle contraction (*systolic dysfunction*) or filling (*diastolic dysfunction*) or both. This is left ventricular failure (LVF or left heart failure) when it affects the left ventricle; right ventricular failure (RVF or right heart failure) when it affects the right ventricle; and congestive heart failure when the failure affects both ventricles. This syndrome can be serious, with life-threatening and often disabling consequences. Symptoms of heart failure include reduced pumping action of the heart, fatigue, coughing, shortness of breath, swollen ankles and legs, swollen abdomen and more.

NOTE [FÉD 20].–

i) Angina is one of the symptoms of coronary heart disease. Coronary artery disease (CAD) is a common and serious condition, leading to heart attacks, heart failure and death. Coronary artery disease is a narrowing of the arteries of the heart, leading to a reduction in the blood supply necessary for heart function. Knowing how to recognize angina is very important, as coronary artery disease can be effectively treated to avoid complications. Angina manifests itself as a deep pain in the middle of the chest, with an intense, agonizing sensation of tightness. This pain occurs when the victim of angina exerts themselves (walking fast, uphill, against the wind) and forces them to stop. It is relieved in less than 2 min by resting and/or taking a nitroxide tablet or spray. This pain is a direct consequence of the heart's suffering, as it no longer receives sufficient blood due to narrowing of the coronary arteries. More rarely, pain occurs after meals, during digestion, or at rest, at night. It can radiate to the left arm, the forearm (sometimes with a sensation of tightness in the wrist), the outer edge of the hand, the neck, the lower jaw, sometimes the right arm, the nape of the neck, the back or the stomach area.

ii) Myocardial infarction is the major complication of angina, but this complication can be avoided or at least delayed for many years if treatment is prescribed early: control of risk factors, medication, revascularization by angioplasty or coronary bypass surgery depending on the degree of artery damage. Angina can begin or worsen abruptly, with attacks that are immediately intense when the patient exerts minimal effort or even at night when resting. This unstable angina pectoris must be considered as a threat of infarction (acute coronary syndrome) and requires emergency hospitalization in a cardiology unit. Conversely, thanks to various treatments, the pain of angina pectoris can disappear, and a completely normal life – active or even sporting (but not just any kind of sport) – is possible. Finally, in some cases, especially if the risk factors for coronary heart disease remain uncontrolled, the disease progresses slowly towards heart failure, with shortness of breath on exertion, and can be complicated by a series of setbacks, including acute coronary syndromes such as myocardial infarction, rhythm disorders and acute pulmonary edema.

2.8. Radiopharmaceuticals used in bone scintigraphy

2.8.1. *Structure of the MDP radiopharmaceutical labeled with metastable technetium-99*

In general, bone scintigraphy is performed by intravenous administration of a radiotracer using a metastable technetium-99 (^{99m}Tc) labeled carrier molecule from the biphosphonate (BP) family. The carrier molecule belongs to the BPs family, characterized by a P-C-P chain. Depending on the possible substitutions on the central carbon atom, several molecules are available for use in bone scintigraphy: MDP (Methylene Di Phosphonate), EHDP (Ethylene Hydroxy Di Phosphonate), HMDP (Hydroxy Methylene Di Phosphonate) and DPD (Dicarboxy Propane Diphosphonate) [AGO 11]. Figure 2.12 shows the special case of the ^{99m}Tc-methyl diphosphonate (^{99m}Tc-MDP) structure [BOT 17].

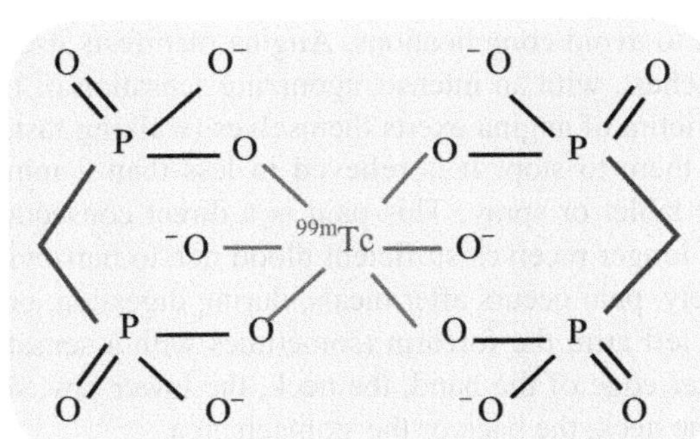

Figure 2.12. *Structure of ^{99m}Tc-methyl diphosphonate (^{99m}Tc-MDP). For a color version of this figure, see www.iste.co.uk/sakho/nuclearphysics3.zip*

2.8.2. *Principle of bone scintigraphy*

For bone scintigraphy using ^{99m}Tc as the radioisotope, the latter is firmly bound to the carrier by a complex radiolabeling reaction (chelation), carried out in a reducing medium. The activity of the injected quantity is of the order of 9 MBq/kg body weight (adult), i.e. around 700 MBq for an average subject weighing 80 kg

[AGO 11]. The injection is usually made at the elbow. The radiopharmaceutical will bind to the skeleton, absorbed into the hydroxyapatite crystals of the mineral bone. Bone uptake is relatively rapid, peaking at 2 h after injection, which is why it is important to allow time between injection and scintigraphy [CHO 09, AGO 11].

The indications and particularities of the bone scan examination are as follows [NAL 14, PEL 18, HEN 19]:

– *Utility*: visualization of bone remodeling, investigation of bone metastases (prostate, breast, bronchial tubes, etc.) in oncology; investigation of pain unexplained by radiology in rheumatology and orthopedics.

– *Indications [HEN 19]*:

- in case of rheumatological or orthopedic pathology (e.g. osteoarthritis, arthritis);

- diagnosis of certain fractures that are difficult to detect by X-ray;

- diagnosis of certain bone infections (osteomyelitis, spondylitis, etc.);

- in the case of cancer (particularly prostate cancer or, more rarely, breast cancer), bone scintigraphy is indicated for the detection of metastases;

- *when a general practitioner or specialist suspects a pathology, they may prescribe a bone scan to rule out cancer or fracture diagnoses.*

– *Contraindication*: pregnancy or breast-feeding. Breast-feeding should be discontinued for 24 h [NAL 14]. Report pregnancy or delayed menstruation to the doctor.

– *Radiopharmaceutical used*: biphosphonates labeled with metastable technetium-99.

– *Patient preparation*: no fasting required as there is no risk to the patient as bone scintigraphy is not dangerous: the quantity of radioactive product injected is extremely small, which is too small to expose the patient to any health risk [PEL 18].

– *Procedure and specifics [NAL 14].*

– Total exam time: approx. 3 h.

– Injections: 1 IV injection.

– Passage under the γ-camera: 1 pass (not systematic, depending on the indication) lasting 10 min, followed by a second pass lasting 20–30 min, 2–3 h after injection.

– Dynamic acquisitions centered under camera from the first minutes after injection (not systematic), followed by whole-body scan acquisitions 2–3 h after radiotracer injection. Images may be supplemented by tomoscintigraphic acquisitions coupled with a CT scan, depending on the indication and abnormalities found during previous acquisitions.

Details of the bone scan procedure are as follows [PEL 18].

To begin with:

– *Step 1*: the radiotracer is injected into a vein in the patient's arm or hand.

– *Step 2*: it is necessary to wait for the substance to cover the entire body and attach itself to the skeleton, i.e. between 2 and 4 h, before moving on to the second step. Depending on the prescription given by the practitioner who ordered the bone scan, the doctor will tell the patient what time they must return to take the images. During this time, the patient has complete freedom of movement, can leave the ward and go for a walk; there are no contraindications and no risks. The only piece of advice is to drink plenty of water to speed up the subsequent elimination of the radioactive tracer.

Moving on:

– *Step 3*: back in the ward, the patient is invited to empty their bladder before proceeding with the actual bone scan.

– *Step 4*: the patient lies on a scintigraphy table, over which the γ-camera passes (the patient does not enter a tube, as in a CT scan). The camera is made up of sensors sensitive to the radioactive radiation of the injected isotope. The doctor or technician can thus visualize the patient's skeleton and observe the areas where the radioactive isotope is most concentrated.

Without prior preparation, the examination takes approximately 15–30 min, depending on the number of images to be taken. At the end of the examination, the injected radiopharmaceutical is eliminated through urination in the hours following the examination. Images are acquired using a γ-camera that can be coupled to a scanner. Anatomical images are thus obtained in addition to the metabolic images from the scintigraphy performed.

Finally, let us note the precautions to be taken after the examination. It is advisable to do the following after the scan [PEL 18]:

– drink plenty of water to eliminate the product;

– wash hands regularly;

– avoid close (less than 1 meter) and prolonged (more than 30 min) contact with young children or pregnant women for the rest of the day.

If the patient needs to be hospitalized in the days following the bone scan, they should inform the department that they have undergone this test.

Some places have radioactivity detectors (airports, certain border crossings, etc.), which the patient may activate several hours after the examination. A certificate indicating the radioactive element administered can be provided on request.

In addition, the risk of allergy is minimal. As the injected substance has no side effects, the patient is perfectly fit to drive a vehicle.

VOCABULARY CORNER.–

– *Miction*: miction (from the latin *mingere*, "to urinate"), the action of urinating refers to the elimination of urine by emptying the bladder. Depending on fluid intake and external humidity and temperature conditions, a healthy person produces around 1–1.5 liters of urine (diuresis) in 24 h, eliminated two to six times a day. We speak of anuria if no urine is eliminated, or if less than 100 ml is eliminated in 24 h.

– *Arthritis*: arthritis and osteoarthritis are two different diseases in the rheumatism family. *Arthritis* is an inflammation of the joint, resulting in the secretion of quinines, substances that gradually destroy the joint. It mainly causes pain when resting (*especially at night*), which may diminish with physical activity or a morning "shake-off". It mainly affects the joints of the hands and feet. *Osteoarthritis*: a so-called "mechanical" pathology: articular cartilage (*resistant, elastic and supple tissue covering the surface of the bones in the joints*) gradually deteriorates and eventually disappears, leaving the bones exposed against each other. While osteoarthritis is a disease that can be linked to age, there are also factors favoring its onset in younger people (one in three patients is under 40). The joints most affected are those of the knee, hip and spine (neck in particular). Unlike arthritis, osteoarthritis tends to cause pain on movement, which diminishes with rest. The deformation of the joint is said to be "cold", as it is not accompanied by any sign of local inflammation and is not visible [BOU 20].

– *Osteomyelitis*: infection of the bone, essentially of bacterial origin. This pathology can affect anyone, but is more common in children and the elderly, particularly in cases of chronic pathology. The main symptoms are intense bone pain, high fever, functional impotence and altered general condition [CHA 21].

– *Spondylitis*: inflammation of the spinal column due to infection, injury or aging.

APPLICATION 2.5.–

A bone scan is performed in a nuclear medicine department by injecting a patient with a solution of metastable technetium-99 at 2 p.m. The initial activity of ^{99m}Tc at the time of injection is 555 MBq. At the end of the examination, the residual activity of ^{99m}Tc is 63% of its initial activity. Determine the time at which the examination is completed.

ANSWER.–

From Table 2.1, we derive the radioactive half-life of ^{99m}Tc: $T = 6.0$ h. Let us determine the examination time. We have:

$$A = A_0 e^{-\ln 2t/T} \Rightarrow t = T\ln(A_0/A)/\ln 2 \qquad [2.19]$$

ANSWER.–

$$t = 6.0 \times \ln(1/0.63)/\ln 2 = 3.999 \text{ h} \approx 4.0 \text{ h}$$

The exam starts at 2 p.m. and lasts 4 h. It therefore ends at 6 p.m.

APPLICATION 2.6.–

A 5 mL vial of the radiopharmaceutical ^{99m}Tc-HDP (^{99m}Tc-hydroxydiphosphonate) is 944.2 MBq at 10 a.m. We wish to carry out a bone scan using this solution. What volume should be injected at 12 p.m.? (Dosage = 630 MBq).

ANSWER.–

Let us determine the activity of the radiopharmaceutical at 12 p.m. The initial activity $A_0 = 944.2$ MBq at $t_0 = 10$ a.m. The activity of the radiopharmaceutical at a later time t is:

$$A(t) = A_0 e^{-\ln 2(t-t_0)/T} = A_0 e^{-\ln 2(t-10)/T} \qquad [2.20]$$

At $t = 12$ p.m., we obtain ($T = 6.0$ h for ^{99m}Tc):

$$A(12) = 944{,}2\, e^{-\ln 2(12-10)/6} = 749.4 \text{ MBq}.$$

The vial contains 5 mL. The activity per milliliter of solution is 749.4 MBq/5 mL = 149.9 MBq/mL.

For a dose of 630 MBq, the volume of solution to be injected is equal to 630/149.9 = 4.2 mL.

2.9. Radiopharmaceuticals used in renal scintigraphy

2.9.1. *Structures of DMSA and MAG$_3$ radiopharmaceuticals labeled with metastable technetium-99*

Renal scintigraphy is an examination used to visualize renal activity and detect any malformations of the kidney or elimination pathways (urethra, bladder). It is performed by injecting a radioactive product into the venous network.

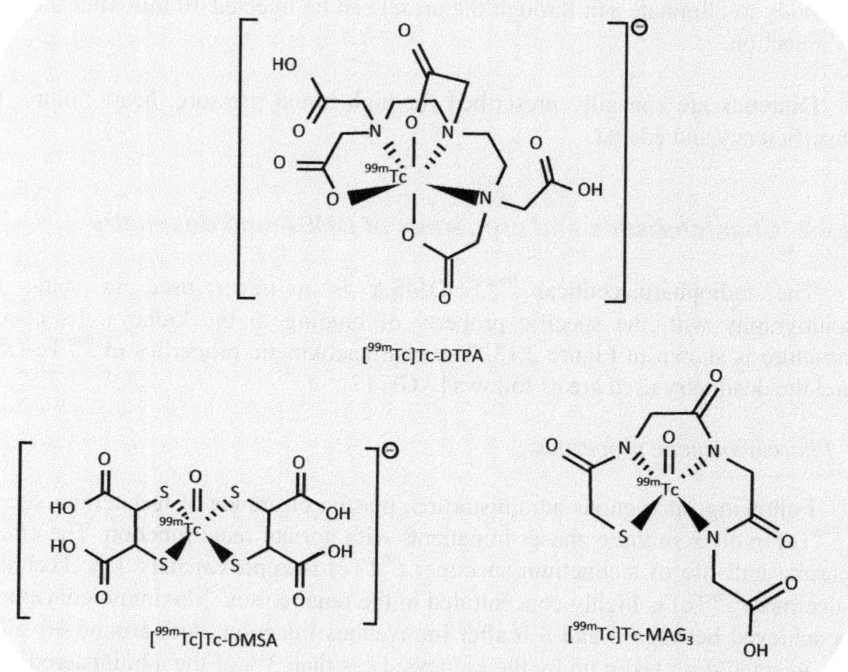

Figure 2.13. *Chemical structures of ^{99m}Tc-DMSA (^{99m}Tc-dimercapto succinic acid), ^{99m}Tc-DTPA (^{99m}Tc-diethylene triamine pentaacetic acid) and ^{99m}Tc-MAG3 (^{99m}Tc-mercapto acetyl triglycine) (source: [RAT 19])*

The examination will show this product collecting in the kidney and then being eliminated by the natural route. It thus reveals obstacles to elimination, reflux, malformations or scars, and assesses kidney function [DEL 19]. This scintigraphy uses dimercapto succinic acid (DMSA) and *mercapto acetyl triglycine* (MAG_3) labeled with metastable technetium-99 (^{99m}Tc-DMSA and ^{99m}Tc-MAG3) as radiopharmaceuticals. The chemical structures of these RPDs are shown in Figure 2.13.

There are two types of renal scintigraphy:

– static renal scintigraphy with DMSA, which binds within the renal parenchyma and enables assessment of the morphology of the two kidneys and detection of renal scars secondary to infectious phenomena in particular;

– dynamic MAG_3 renal scintigraphy to assess the relative function of each kidney. To accelerate emptying of the urinary tract, a diuretic (a drug that enables the body to eliminate salt through the urine) can be injected 10 min after the MAG_3 IV injection.

Diuretics are specially prescribed for high blood pressure, heart failure, renal insufficiency and edema.

2.9.2. Characteristics and properties of DMSA and dosimetry

The radiopharmaceutical ^{99m}Tc-DMSA is a tracer used in static renal scintigraphy with the specific property of binding in the kidneys. Its chemical structure is shown in Figure 2.13. The pharmacokinetic properties of ^{99m}Tc-DMSA and the dosimetry used are as follows [AGE 17]:

– *Pharmacokinetic properties*:

Following intravenous administration, plasma clearance of technetium succimer (^{99m}Tc) evolves in three phases in patients with normal renal function. The effective plasma half-life of technetium succimer (^{99m}Tc) is approximately 1 h. Technetium succimer (^{99m}Tc) is highly concentrated in the renal cortex. Maximum concentration is achieved between 3 and 6 h after intravenous injection, with around 40–50% of the injected dose taken up by the kidneys. Less than 3% of the administered dose is found in the liver. Nevertheless, hepatic levels may be significantly increased and renal uptake reduced in patients with renal insufficiency.

– *Dosimetry*:

As we already know, technetium (^{99m}Tc) decays by emitting mainly 140 keV γ-radiation, with a half-life of 6 h, to give technetium (^{99}Tc), which can be considered stable. The effective dose, after intravenous administration of an activity of 120 MBq, is 1.06 mSv (in a 70 kg individual).

Table 2.13 shows the radiation doses absorbed by patients in the case of ^{99m}Tc-DMSA scintigraphy [TOU 01]. It should be noted that the dose received by the patient depends on their age and the organ explored.

^{99m}Tc	Technetium succimer (DMSA)				
	Intravenous injection				
	Absorbed dose per unit of activity administered				
	(µGy/MBq)				
Organs	**Adult**	**15 years**	**10 years**	**5 years**	**1 year**
Kidneys	180	220	300	430	760
Bladder wall	18	23	29	31	57
Spleen	13	17	26	38	61
Ovaries	3.5	4.7	7.0	1.1	1.9
Testes	1.8	2.4	3.7	5.3	10
Uterus	4.5	5.6	8.3	11	19
Effective dose (µSv/MBq)	8.6	11	15	17	28

Table 2.13. *Absorbed doses in ^{99m}Tc-DMSA static renal scintigraphy. For a color version of this table, see www.iste.co.uk/sakho/nuclearphysics3.zip*

Note that the activity for children should be calculated according to their body mass, using the following formula [AGE 17]:

Pediatric activity (MBq) = [Adult activity (MBq) × Child's body mass (kg)]/70 [2.21]

In some cases, calculation based on body surface area is more appropriate:

Pediatric activity (MBq) = [Adult activity (MBq) × Child's body surface area (m^2)]/1.73 [2.22]

NOTE ON RENAL FAILURE [GUÉ 19].–

Renal failure corresponds to a change in the functioning of the kidneys, which no longer perform their functions. There are two types of renal failure: chronic renal failure and acute renal failure. The former poses a real problem in terms of public health: its incidence is increasing, which is complicated by high morbidity.

– *Chronic renal failure* is defined as kidney dysfunction lasting at least three months, characterized by a GFR (glomerular filtration rate) of less than 60 mL/min relative to body surface area. In this pathology, the kidney becomes progressively fibrotic. Depending on the GFR, kidney failure may be classified as incipient (mild), moderate, severe or terminal.

– *Acute renal failure* occurs in a healthy kidney. It is a reversible phenomenon linked to accidents such as severe dehydration, shock or medication such as nonsteroidal anti-inflammatory drugs.

Diabetes and hypertension are the two major causes of kidney failure.

VOCABULARY CORNER.–

– *Glomerular filtration rate (GFR)*: is the volume of liquid filtered by the kidney per unit of time. It is a value that quantifies the activity of the kidney. GFR is classically estimated by serum creatinine, calculation of creatinine clearance from 24 h urine collection or by the Cockcroft-Gault formula. This formula (see [DON 76]) was proposed by Donald W. Cockcroft and Henry Gault in 1976 and is used to estimate creatinine clearance in adults. Since creatinine clearance is very close to glomerular filtration rate, it provides information on the state of renal function. Creatinine analysis provides information on kidney function and muscle mass. Creatinine is produced when the kidneys break down the muscle protein creatine. A high creatinine level is often a sign of kidney failure. A low creatinine level may be a sign of myopathy (severe muscle atrophy) [ODO 18].

– *Steroidal anti-inflammatory drugs or corticoids*: these are drugs used to treat inflammation, and sometimes certain allergic reactions. *Nonsteroidal anti-inflammatories: the best-known are aspirin and ibuprofen. Steroidal anti-inflammatories, on the contrary, are corticoids, derived from cortisol and cortisone, and have a more powerful effect [FER 19].*

– *Cortisol*: the hormone produced by the adrenal glands. It plays an essential role in balancing blood glucose and releasing sugar from the body's reserves in response to an increased demand for energy.

– *Cortisone*: *a hormone naturally secreted by the adrenal glands in the cortex. It is a hormone essential to life, with anti-inflammatory properties and regulates sugar, protein and carbohydrate metabolism*. Natural cortisone can also be manufactured in a laboratory for prescription use. It mimics the cortisone naturally produced by the adrenal glands. It is administered to people with adrenal insufficiency, who can no longer secrete this hormone [GUÉ 19].

– *Creatinine*: *an organic component naturally produced by the body* synthesized by the kidneys, liver and pancreas and stored in muscle fibers. It *comes from our diet* and can also be supplemented.

– *Creatinine*: normal metabolic waste produced by the body. It is the result of creatine breakdown, eliminated mainly by the kidneys. Its level in the body depends on renal elimination capacity and muscle mass.

– *Gland*: one or more cells whose function is to secrete substances that are stored and then excreted.

– *Adrenal glands*: located just above each kidney, these glands produce numerous hormones such as cortisol and adrenaline, which are necessary for the body to function properly.

2.9.3. ^{99m}Tc-DMSA static scintigraphy procedure

Let us summarize the main information on ^{99m}Tc-DMSA static renal scintigraphy [NAL 14, AGE 21a, AGE 21b, CEN 21b].

– *Utility*: to observe the morphology of the kidney.

– *Indications*: after reconstitution with sodium pertechnetate (^{99m}Tc) injection, ^{99m}Tc-DMSA can be used for the purpose of:

- morphological studies of the renal cortex;
- assessment of the functional value of each kidney;
- localization of renal ectopy.

– *Contraindications*: pregnancy [NAL 14].

– *Patient preparation*: no fasting required.

– *Procedure and specifics*:

- total examination time: approximately 6 h;
- injections: 1 IV injection into an arm vein;
- passage under the γ-camera: one pass of 10–30 min or even 40 min;
- images are taken 6 h after IV injection: DMSA fixation is maximal after 4–5 h. This is the time to wait after intravenous injection. A planar scintigraphic image is then taken, with the γ-camera centered on the patient's back.

2.9.4. *MAG₃ characteristics and properties and dosimetry*

As mentioned above, the radiopharmaceutical $^{99m}Tc\text{-}MAG_3$ (^{99m}Tc-*mercapto acetyl triglycine*) is used in dynamic renal scintigraphy to assess kidney function. The pharmacokinetic properties of $^{99m}Tc\text{-}MAG_3$ and the dosimetry used are as follows [AGE 09].

Pharmacokinetic properties

After intravenous injection, (^{99m}Tc) MAG_3 is rapidly eliminated from the blood by the kidney. (^{99m}Tc) MAG_3 is relatively highly bound to plasma proteins (66%). When renal function is normal, 70% of the administered activity is excreted after 30 min, and over 95% after 3 h. These percentages are modified in the case of renal and/or urological pathologies. Excretion is predominantly by tubular secretion, with glomerular filtration accounting for only 10% of total clearance. It should be noted that tubular secretion enables the elimination in urine of undesirable or excess substances that have been insufficiently filtered by the glomerulus.

Dosimetry

The effective dose resulting from the administration of a 200 MBq activity to a 70 kg individual is 1.4 mSv, when the bladder is emptied 2 h after administration. The radiation doses absorbed by patients are shown in Table 2.14 [TOU 01].

	Technetium mertiatide (MAG_3)				
^{99m}Tc	**Intravenous injection/Normal renal function Absorbed dose per unit of activity administered (µGy/MBq)**				
Organs	**Adult**	**15 years**	**10 years**	**5 years**	**1 year**
Bladder wall	110	140	170	180	320
Colon wall	3.4	4.3	5.9	6.0	9.8
Kidneys	3.4	4.2	5.9	8.4	15
Ovaries	5.4	6.9	8.7	8.7	14
Testes	3.7	5.3	8.1	8.7	16
Uterus	12	14	19	19	31
Effective dose (µSv/MBq)	**7.4**	**9.4**	**12**	**12**	**22**

^{99m}Tc	Pathological kidney function Absorbed dose per unit of activity administered				
	(µGy/MBq)				
Organs	**Adult**	**15 years**	**10 years**	**5 years**	**1 year**
Bladder wall	83	110	130	130	230
Colon wall	14	17	24	3.4	59
Kidneys	3.4	44	6.1	6.9	11
Ovaries	4.9	63	8.1	8.7	14
Testes	3.4	47	7.1	7.8	14
Uterus	10	12	16	16	27
Effective dose (µSv/MBq)	**6.3**	**8.3**	**10**	**11**	**19**
^{99m}Tc	Acute unilateral renal blockage Absorbed dose per unit of activity administered				
	(µGy/MBq)				
Organs	**Adult**	**15 years**	**10 years**	**5 years**	**1 year**
Kidneys	200	240	330	470	810
Bladder wall	56	71	91	93	170
Adrenal glands	11	14	22	32	55
Ovaries	3.8	5.1	7.1	9.2	15
Testes	2.0	2.9	4.5	5.0	9.8
Uterus	7.2	8.7	12	13	22
Effective dose (µSv/MBq)	**10**	**13**	**17**	**22**	**38**

Table 2.14. *Absorbed doses in dynamic renal scintigraphy with ^{99m}Tc-MAG$_3$. For a color version of this table, see www.iste.co.uk/sakho/nuclearphysics3.zip*

2.9.5. *Characteristics and properties of DTPA and dosimetry*

As mentioned above, the radiopharmaceutical ^{99m}Tc-DTPA (^{99m}Tc-diethylene triamine pentaacetic acid) is used in dynamic renal scintigraphy. It is eliminated by the kidneys and is used to assess kidney function. The pharmacokinetic properties of ^{99m}Tc-DTPA and the dosimetry used are as follows [AGE 19].

– *Pharmacokinetic properties*:

After intravenous injection, technetium pentetate (^{99m}Tc) (DTPA) is rapidly distributed in the extracellular volume. Less than 5% of the injected dose is bound to plasma proteins. A negligible percentage of the injected activity of technetium pentetate (^{99m}Tc) is bound to red blood cells. Technetium pentetate (^{99m}Tc) (DTPA) does not cross the normal blood–brain barrier. It diffuses only slightly into breast milk. Plasma clearance is multi-exponential and includes a particularly rapid phase.

The complex remains stable in vivo: over 98% of the urinary activity is in chelate form. Approximately 90% of the injected dose is eliminated in the urine within the first 24 h, mainly by glomerular filtration. No renal intra-parenchymal retention of the product has been demonstrated. Renal clearance may be reduced in patients with kidney disease. In individuals with edema or ascites, the distribution of technetium pentetate (^{99m}Tc) in the extracellular volume may be impaired.

– *Dosimetry*:

For technetium pentetate, the effective dose for a 70 kg individual resulting from the intravenous administration of an activity of 740 MBq to a patient with normal renal function is 4.7 mSv [AGE 19]. The radiation doses absorbed by patients, according to ICRP 80, are given in Table 2.15 [TOU 01] for normal renal function [AGE 19].

^{99m}Tc	Technetium pentetate (DTPA)				
	Intravenous injection				
	Absorbed dose per unit of activity administered (µGy/MBq)				
Organs	**Adult**	**15 years**	**10 years**	**5 years**	**1 year**
Bladder wall	62	78	97	95	170
Kidneys	3.9	4.7	6.7	9.6	17
Colon wall	3.0	3.8	5.4	6.4	11
Ovaries	4.2	5.3	6.9	7.8	13
Testes	2.9	4.0	6.0	6.9	13
Uterus	7.9	9.5	13	13	22
Effective dose (µSv/MBq)	**5.1**	**6.4**	**8.5**	**9.2**	**16**

Table 2.15. *Absorbed doses in dynamic renal scintigraphy with ^{99m}Tc-DTPA for normal renal function. For a color version of this table, see www.iste.co.uk/sakho/nuclearphysics3.zip*

For patients with impaired renal function, the radiation doses absorbed are given in Table 2.16 [AGE 19].

^{99m}Tc	Technetium pentetate (DTPA)				
	Intravenous injection				
	Absorbed dose per unit of activity administered (μGy/MBq)				
Organs	**Adult**	**15 years**	**10 years**	**5 years**	**1 year**
Bladder wall	22	27	40	58	110
Kidneys	7.9	9.6	14	20	34
Upper colon wall	4.4	5.6	8.1	13	22
Lower colon wall	4.7	6.2	9.6	14	25
Ovaries	4.9	6.3	9.4	14	24
Testes	3.3	4.5	6.9	11	20
Uterus	6.3	7.5	11	17	29
Effective dose (μSv/MBq)	**4.8**	**6.1**	**9.0**	**13**	**24**

Table 2.16. *Absorbed doses in ^{99m}Tc-DTPA dynamic renal scintigraphy for impaired renal function. For a color version of this table, see www.iste.co.uk/sakho/nuclearphysics3.zip*

2.9.6. ^{99m}Tc-TDPA and ^{99m}Tc-MAG$_3$ dynamic scintigraphy procedure

Let us summarize the main information relating to dynamic renal scintigraphy with ^{99m}Tc-TDPA [NAL 14, AGE 19, CEN 21b].

– *Utility*: estimation of renal functional distribution; exploration of urinary excretion; analysis of drainage from renal excretory cavities.

– *Indications*:

After reconstitution with an injectable solution of sodium pertechnetate (^{99m}Tc), ^{99m}Tc-TDPA can be used for:

- the study of perfusion, renal function and the urinary tract;

- the determination of glomerular filtration rate.

– *Patient preparation*: no fasting necessary. In order to reduce the radiation dose and the accumulation of radioactivity in the bladder, the patient will be asked to drink copiously (two to three glasses of water) and to empty the bladder as often as possible in the hours following administration.

– *Procedure and particularities*:

- total exam time: 1 h 30 min;
- injections: one to two IV injections (tracer and furosemide);
- passage under the γ-camera: one pass of 45 min;
- images are taken approximately 40 min after injection.

NOTE [AGE 19].–

– *Diagnostic use of captopril*:

Dynamic renal scintigraphy, first performed under controlled conditions and then repeated 1 h after oral administration of captopril (25–50 mg), reveals variations in the hemodynamics of the kidney with a stenotic artery. Blood pressure should be monitored carefully, as patients with renovascular hypertension are at risk of significant hypotension and impaired renal function following administration of captopril.

– *Diagnostic use of furosemide*:

Intravenous administration of furosemide during dynamic renal scintigraphy accelerates the evacuation of technetium pentetate (^{99m}Tc), which helps in the differential diagnosis between true obstruction and dilatation of the urinary tract.

VOCABULARY CORNER.–

– *Hypotension*: drop in blood pressure. Blood pressure is normal if its value is close to 120/80 mm Hg (the first figure corresponds to systolic pressure, the second to diastolic pressure (maximum/minimum value)).

– *Renovascular hypertension*: increase in blood pressure caused by partial or complete occlusion (closure) of one or more renal arteries or their branches.

– *Stenosis*: this is a permanent narrowing of the caliber of a canal or blood vessel.

– *Mannitol*: mannitol is hexane-1,2,3,4,5,6-hexol ($C_6H_{14}O_6$) and is a polyol.

– *Captopril*: captopril is 1-[(2*S*)-3-mercapto-2-methylpropionyl]-L-proline ($C_9H_{15}NO_3S$). It is a drug used in the treatment of arterial hypertension.

– *Furosemide*: furosemide is 5-(aminosulfonyl)-4-chloro-2-[(2-furanylmethyl) amino] benzoic acid ($C_{12}H_{11}ClN_2O_5S$).

The chemical structures of mannitol, captopril and furosemide are shown in Figure 2.14.

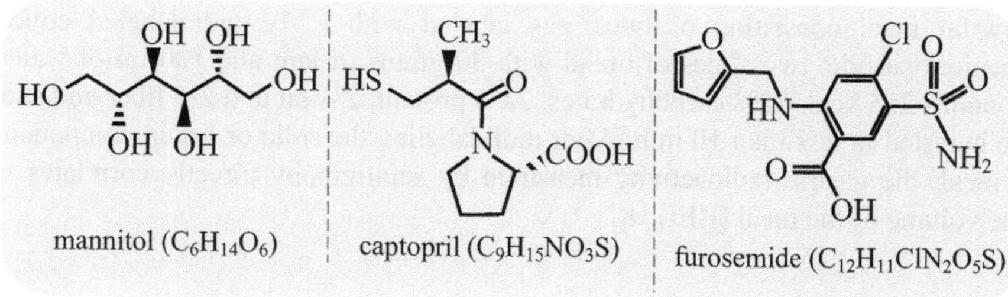

Figure 2.14. *Chemical structures of mannitol (hexane-1,2,3,4,5,6-hexol), captopril (1-[(2S)-3-mercapto-2-methylpropionyl]-L-proline) and furosemide (5-(aminosulfonyl)-4-chloro-2-[(2-furanylmethyl) amino] benzoic acid)*

2.10. Radiopharmaceuticals used in digestive scintigraphy

2.10.1. *Definition*

Digestive scintigraphy, also known as gastric emptying scintigraphy, is used to monitor the passage of a meal through the stomach and to study stomach contractions. This examination is used in the event of digestive disorders, when poor emptying of the stomach (known as gastroparesis) is suspected, for example during diabetes. It can also be used to explore gastro-esophageal reflux or to look for gastroparesis, particularly in diabetic patients [NAL 14]. *Gastroparesis* is a syndrome characterized by delayed gastric emptying in the absence of mechanical obstruction. Patients with gastroparesis may exhibit a wide spectrum of clinical manifestations including nausea, vomiting, postprandial pain and early satiety [BEL 18].

2.10.2. *Standardized consensus meal labeled with metastable technetium-99*

Many types of test meal are used, including liquid, solid and mixed liquid/solid meals. The radioactive label most frequently used for solid components is based on rhenium sulfide nanocolloids radiolabeled with ^{99m}Tc. This isotope has a short half-life of around 6 h, which limits exposure to radiation. Once cooked, it is tightly bound to the egg proteins, so that no radioactive material is absorbed by the gastrointestinal tract. For liquid components, the ^{113}In isotope is more suitable, with a half-life of 90 min. However, it is not approved for routine clinical use. When a dual-head γ-camera is used, emptying of the solid and liquid fractions of the meal can be measured simultaneously.

The standardized meal adopted for gastric scintigraphy is based on consensus recommendations published jointly by the American Neurogastroenterology and Motility Society and the Society of Nuclear Medicine [ABE 08]. This is a specific low-fat meal consisting of two eggs labeled with ^{99m}Tc radiolabeled colloidal rhenium sulfide, two slices of bread with 30 grams of jam and 120 ml of water. It contains 255 kcal, 72% carbohydrates, 24% protein, 2% fat and 2% fiber and should be ingested in less than 10 min. After radiolabeling the solid or liquid component of a meal, the gastric radioactivity measured by scintigraphy directly correlates with the volume of the meal [BEL 18].

2.10.3. *Digestive scintigraphy procedure*

Let us summarize the procedure and particularities of digestive scintigraphy [NAL 14]:

– *Utility*: assessment of the kinetics of gastric emptying of solids after ingestion of a standardized radiolabeled meal: the radioactivity present in the stomach is proportional to the quantity of food present. A slowdown or acceleration in the emptying of solids can be demonstrated.

– *Indications*:

- investigation of gastroparesis, particularly in diabetic patients;

- investigation of a dyspeptic syndrome;

- gastroesophageal reflux.

– *Contraindications*: pregnancy; allergy to foods in the standard meal.

– *Radiopharmaceutical*: standardized meal labeled with ^{99m}Tc (egg white, bread, jam, water, etc.).

– *Patient preparation*: 12 h fast.

– *Procedure and particularities*:

- total exam time: 4 h;

- administration: oral administration of the standardized meal;

- passage under the γ-camera: four passages of 15 min;

- static images centered on the stomach were taken immediately after injection of the meal and at 30 min, 1 h, 2 h, 3 h and 4 h.

3

Radioisotopes Fluorine-18, Metastable Krypton-81 and Iodine-123, 125 and 131 used in Nuclear Medicine

General objective	
Understand the properties and uses of radiopharmaceuticals labeled with fluorine-18 and iodine-127 and of metastable krypton-81 radiotracers and radioiodines-123, 125 and 131 in nuclear medicine.	
Specific objectives	
Understand the main production reaction of radiofluorine-18 for medical use;	Understand the properties of the radiopharmaceuticals [^{123}I] MIBG (MetaIodoBenzylGuanidine) and [^{123}I] ioflupane;
Interpret the decay energy diagram for radiofluorine-18;	Understand the principle of iodine-123 thyroid scintigraphy and dosimetry;
Understand the synthesis mechanism of the radiopharmaceutical 2-deoxy-2-[^{18}F]fluoro-D-glucose (^{18}FDG);	Understand the use of the radiopharmaceutical ^{123}I-ioflupane (Datscan®) in the differential diagnosis of essential tremor and neurodegenerative diseases such as Parkinson's, Alzheimer's and Lewy body dementia;
Understand the benefits of the radiopharmaceutical ^{18}FDG most commonly used in PET scans;	Explain the mechanism of excretion of iodine-123 from the body after an IV injection;
Demonstrate the principle of ^{18}FDG scintigraphy and the dosimetry used;	Understand the value of radioiodine-131 in thyroid scintigraphy;
Explain the mechanism of production of two γ photons after injection of ^{18}FDG into the body;	Understand the ^{131}I radiotracer production chain from the generator ^{130}Te (n, γ) ^{131m,g}Te $\rightarrow$ ^{131}I;
Explain the mechanism by which fluorine-18 is excreted from the body;	Understand the main emissions of iodine-131;
Understand the production line for metastable krypton-81 (^{81m}Kr) tracer using the ^{81}Rb/^{81m}Kr generator;	Interpret the decay diagram for iodine-131;

Understand the properties of metastable krypton-81 used in lung ventilation scintigraphy;	Understand the effect of prolonged decay of iodine-131 on thyroid cancer;
Be familiar with the principle of ^{81m}Kr lung ventilation scintigraphy and the dosimetry used;	Be familiar with the principle of radiotherapeutic treatment of thyroid cancer with iodine-131 and the dosimetry used;
Explain the mechanism of clearance of ^{81m}Kr in the body after inhalation;	Explain the excretion mechanism of iodine-131 from the body;
Know how the ^{81m}Kr stress myocardial scan is performed;	Define brachytherapy;
Understand the role of iodine in the body;	Know the symptoms and diagnosis of prostate cancer;
Understand the benefits of radioiodine-123 in thyroid scintigraphy;	Understand the principle of prostate brachytherapy using iodine-125 implants;
Understand the production chain for radiotracer ^{123}I via generators $^{127}I\ (p, 5n)\ ^{123}Xe$, $^{121}Sb\ (\alpha, 2n)\ ^{123}I$, $^{122}Te\ (d, n)\ ^{123}I$ and $^{122}Te\ (\alpha, 3n)\ ^{123}Xe$;	Interpret the decay energy diagram for iodine-125;
Understand the main emissions of the radiotracer ^{123}I;	Understand the fundamental difference between prostate brachytherapy and prostate chemotherapy.
Prerequisites	
Beta decay;	Isomeric transition;
Electronic capture (EC) process;	Nuclear transmutation reactions;
Auger effect;	Structure of the heart, lungs, brain, kidneys and stomach.
Gamma de-excitation;	

3.1. Properties and uses of the radiotracer ^{18}FDG

Positron-emitting radionuclides can be used in positron emission tomography (PET) on specific labeled molecules. However, none of these radionuclides has led to the development of a commercial product, due to an even shorter half-life. This is the case, for example, with radiocarbon-11, which has a very short half-life of about 20 min. Today, fluorine-18 is still unrivalled in the manufacture of radiopharmaceuticals for use in PET. For this reason, we will study the properties and use of this radioisotope in detail in this section.

A cancer cell is known as a glucose-hungry tumor. It therefore consumes more glucose than a normal cell. To differentiate between a tumor that consumes a lot of glucose and a healthy cell, a radiopharmaceutical of interest is prepared, mainly based on radiofluorine-18.

3.1.1. *Choice of radiofluorine-18 in positron emission tomography*

Among the positron-emitting radionuclides that can be used in PET, the radioisotopes ^{13}N, ^{11}C, ^{18}F and ^{15}O are mainly distinguished [DED 04, GUI 17, LEB 23]. The main physical characteristics of these radioisotopes are summarized in Table 3.1.

Isotopes	^{11}C	^{13}N	^{15}O	^{18}F
Maximum kinetic energy of β^+ (MeV)	0.98	1.19	1.72	0.63
Most probable kinetic energy of β^+ (MeV)	0.39	0.49	0.73	0.25
Maximum free path in the water (mm)	3.9	5	7.9	2.3
Mean free path in the water (mm)	1.1	1.5	2.7	0.6
Radionuclide half-life (min)	20.4	10.0	2.1	109.8

Table 3.1. *Physical characteristics of the main positron-emitting isotopes used in PET. The kinetic emission energy of positrons (β^+) conditions the free path of this particle before the annihilation reaction [DED 04]. Mean free path: the average distance traveled by the positron during the decay of the radionuclide before the annihilation process. For a color version of this table, see www.iste.co.uk/sakho/nuclearphysics3.zip*

The radioisotopes ^{13}N, ^{11}C and ^{15}O can therefore only be produced in the few research centers that produce these nuclides on site using a cyclotron. The relatively short half-life of ^{11}C, ^{13}N and ^{15}O prevents the use of molecules labeled by these isotopes at a distance from the production site. Only ^{18}F has a period of about 110 min compatible with routine clinical use. This requires the delivery of the nuclear medicine department several times a day in radiopharmaceuticals marked with ^{18}F [LEB 23]. In addition, its positrons have a relatively low maximum energy (0.63 MeV), which results in a fairly short average path through the tissues before annihilation in water or tissues, and therefore a better image resolution [GUI 17].

3.1.2. *Obtaining radiofluorine-18*

Depending on the type of radiolabeling envisaged, fluorine-18 will be used either in electrophilic form with molecular fluorine $[^{18}F]F_2$ or in nucleophilic form with fluoride ions $[^{18}F]F^-$. The "electrophilic" fluorine-18 is produced by irradiation of F_2-enriched neon gas (2%), according to the nuclear reaction ^{20}Ne (d, α) ^{18}F, and is obtained as gaseous $[^{18}F]F_2$. Maximum radiochemical yields are limited to 50% (in difluorine $[^{18}F]F_2$, only one of the two fluorine atoms is labeled and incorporated into the reactant or product).

However, "nucleophilic" fluorine-18 is most often produced by irradiation of a target of oxygen-18 enriched water (> 95%), according to the $^{18}O\ (n,\ p)^{18}F$ nuclear reaction. It is obtained as a fluoride anion $[^{18}F]F^-$ in aqueous solution with a high specific radioactivity (> 185 GBq/µmol). These reactions are those that allow acceptable production by small cyclotrons for medical use [DOL 06, VAL 09, AUS 11, GUI 17]. Radiochemical yields of 60–65% can be obtained.

APPLICATION 3.1.–

Explain the nuclear reactions $^{20}Ne\ (d,\ \alpha)^{18}F$ and $^{18}O\ (p,\ n)^{18}F$ to produce radiofluorine ^{18}F, replacing the particles d, α, p and n and the nuclides ^{20}Ne, ^{18}O and ^{18}F with their symbols.

ANSWER.–

These are induced nuclear reactions which we studied in Chapter 4 of Volume 1 [SAK 22]. They are of the type $X\ (a,\ b)\ Y$ and are generally written as:

$$a + X \rightarrow b + Y \qquad [3.1]$$

In this equation, X is the *target nucleus*, Y is the *residual nucleus*, a is the *incident particle* and b is the *emergent particle*. We then obtain (d: deuteron; α: helium-4 nucleus; p: proton and n: neutron):

– Reaction $^{20}Ne\ (d,\ \alpha)^{18}F$

$$^{2}_{1}H + ^{20}_{10}He \rightarrow ^{4}_{2}He + ^{18}_{9}F \qquad [3.2]$$

– Reaction $^{18}O\ (p,\ n)^{18}F$

$$^{1}_{1}H + ^{18}_{8}O \rightarrow ^{1}_{0}n + ^{18}_{9}F \qquad [3.3]$$

Note that the nuclear reaction $^{20}Ne\ (d,\ \alpha)^{18}F$ only allowed the production of fluorine in gaseous form, which could only be recovered from targets with another gas as a carrier (F_2). The use of a toxic carrier gas such as F_2 requires more equipment and qualified personnel in order to be handled. This method is almost no longer used [AND 18].

However, the $^{18}O(n,\ p)^{18}F$ [3.3] reaction most commonly used in radiochemistry produces large quantities of fluorine-18 with high specific activity (from 111 to 370 GBq/µmol) [MED 14]. In addition, this reaction is carried out in a cyclotron with protons accelerated to very high energies above 16 MeV [INS 13a]. Reaction [3.2] is also used by deuteron bombardment with energies of > 8 MeV.

The ^{18}F radioisotope decays without emission of γ [AUS 11], mainly 96.7% (≈ 97%) by position emission and 3.1% (≈ 3%) by electron capture (EC), followed by emission of low-energy X-rays [INS 13a, POT 14]. After slowing down in matter, the positrons produced annihilate with electrons. This produces two γ photons (194%), each with 511 MeV energy. The decay diagram of ^{18}F summarizing the radiophysical properties of ^{18}F is shown in Figure 3.1.

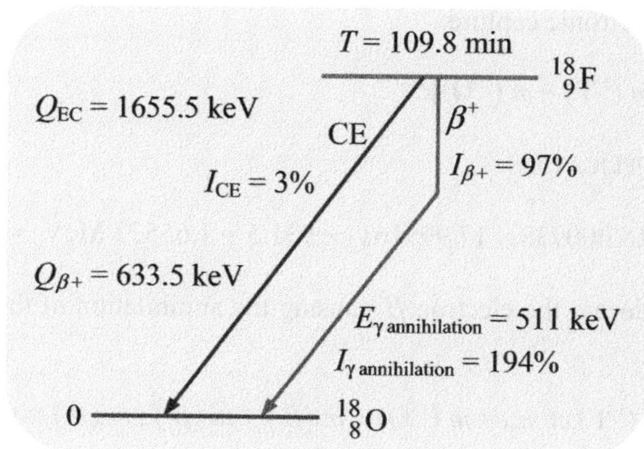

Figure 3.1. *Diagram of the decay of radiofluorine-18. For a color version of this figure, see www.iste.co.uk/sakho/nuclearphysics3.zip*

The decay product is oxygen-18 (stable). The decay process can be summarized as follows:

$$^{18}_{9}F \xrightarrow[\beta^+\ (97\%)]{CE\ (3\%)} {}^{18}_{8}O \qquad [3.4]$$

The radiophysical properties of fluorine-18 are determined by its half-life, its main emissions (particles β^+ and photons γ of annihilation), its decay and annihilation energies and its parentage.

APPLICATION 3.2.–

Retrieve the values of the electron capture energies (Q_{EC}) and decay β^+ ($Q_{\beta+}$) of radiofluorine-18.

Data: m (^{18}F) = 18.0009380 u; m (^{18}O) = 17.9991610 u; $m_{\beta+} = m_0$ = 5.4857990946×10^{-4} u; 1 u = 931.5 MeV/c^2.

ANSWER.–

By definition (m_{be} is the mass of the particles before the reaction and m_{af} denotes the mass of the particles obtained after the reaction):

$$Q = (m_{be} - m_{af})c^2 \qquad [3.5]$$

– For the electronic capture:

$$Q_{CE} = [m\,(^{18}F) - m\,(^{18}O)]c^2 \qquad [3.6]$$

NUMERICAL APPLICATION.–

$$Q_{CE} = [18.000938 - 17.999161] \times 931.5 = 1.65527 \text{ MeV } \approx 1655.\,3 \text{ keV}$$

– For beta decay, the electron β^- causing the annihilation of the positron is taken into account:

$$m_{be} = m\,(^{18}F) \text{ et } m_{af} = m\,(^{18}O) + m\,(\beta^+) + m\,(\beta^-) : m\,(\beta^+) = m\,(\beta^-) = m_0.$$

$$Q_{\beta+} = [m\,(^{18}F) - m\,(^{18}O) - 2m_0]c^2 \qquad [3.7]$$

NUMERICAL APPLICATION.–

$$Q_{\beta+} = [18{,}000938 - 17{,}999161 - 2 \times 5.4857990946 \times 10^{-4}] \times 931.5 = 0{,}63327 \text{ MeV } \approx 633.3 \text{ keV}$$

3.1.3. *Synthesis of the radiopharmaceutical ^{18}FDG*

Among all ^{18}F-labeled radiopharmaceuticals, ^{18}FDG is considered as the reference radiotracer for in vivo cancer imaging. This preference is justified by the fact that deoxyglucose is an analogue of glucose, which competes with the latter in its uptake by macrophages and other glucose-consuming cells [DEB 08, GUI 17].

VOCABULARY CORNER.–

Glucose: sugar with the crude formula $C_6H_{12}O_6$. This is naturally present in all living things on Earth and is their most important source of energy. It is found in quantities in fruits, vegetables, honey, etc.

Deoxyglucose: sugar corresponds to 2-deoxy-d-glucose with the chemical formula $C_6H_{12}O_5$. We see that it is a glucose molecule whose 2-hydroxyl group (OH) is replaced by a hydrogen atom (H).

Macrophage: a kind of white blood cell that surrounds and kills bacteria or microorganisms, absorbs foreign substances, removes old or damaged cells and helps stimulate the immune system.

In formal terms, ^{18}FDG is obtained by replacing a hydroxyl group (OH) in a glucose molecule with a radioactive fluorine atom ^{18}F (Figure 3.2).

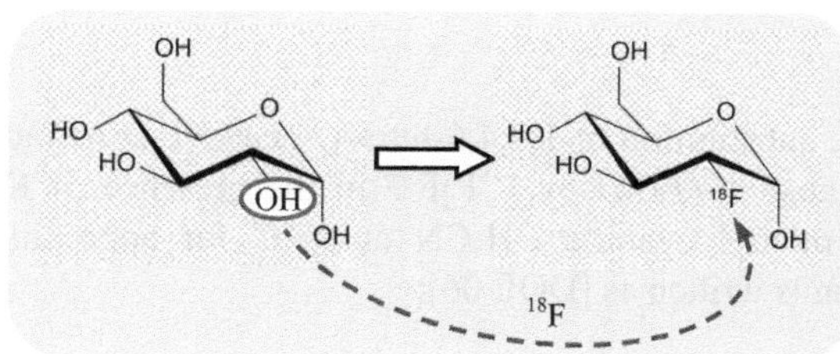

Figure 3.2. *Formal diagram for obtaining 2-deoxy-2-[^{18}F]fluoro-D-glucose (^{18}FDG) by replacing a hydroxyl group (OH) in a glucose molecule with a radioactive fluorine-18 (^{18}F) atom. For a color version of this figure, see www.iste.co.uk/sakho/nuclearphysics3.zip*

In practice, [^{18}F]FDG is synthesized by nucleophilic substitution of 1,3,4,6-tetra-*O*-acetyl-2-*O* trifluoromethanesufonyl-β-D-mannopyranose 1 by [^{18}F]F^- in the presence of a cryptant, Kryptofix 2.2.2. The reaction is carried out in acetonitrile at 80°C for approximately 4 min, after which the hydroxyl functions can be deprotected either in an acid medium (HCl) or in a basic medium (NaOH); this leads to the desired product in good radiochemical yields of around 60–65% [VAL 09]. Kryptofix 2.2.2 is a crystal composed of potassium ion K^+ and fluoride ion $^{18}F^-$. Its formula is shown in Figure 3.3 [AND 18].

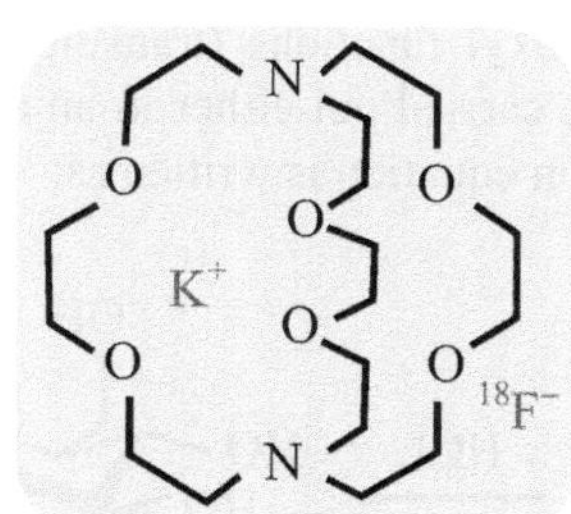

Figure 3.3. *Chemical structure of the Kryptofix 2.2.2 cryptant. For a color version of this figure, see www.iste.co.uk/sakho/nuclearphysics3.zip*

The Kryptofix 2.2.2 cryptant makes it possible to improve the solubility of $^{18}F^{-}$ ions in the form $[^{18}F]F^{-}$ in the reaction medium and to increase their nucleophilicity (attractiveness to positive sites).

As mentioned above, ^{18}FDG is synthesized in two main stages [VAL 09, AND 18, HER 19]:

First stage

Nucleophilic substitution of 1,3,4,6-tetra-*O*-acetyl-2-*O* trifluoromethanesufonyl-β-D-*m*annopyranose 1 *(TTM)* by $[^{18}F]F^{-}$ in the presence of Kryptofix 2.2.2 and acetonitrile or methyl cyanide CH_3CN at 85°C for approximately 4 min. The reaction equation is written as [DOL 06]:

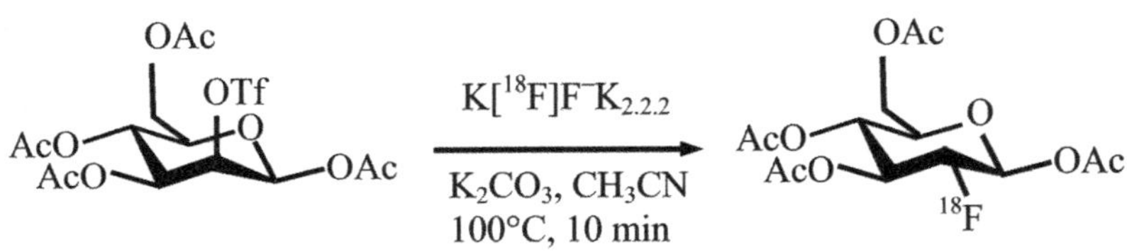

[3.8]

TTM

The acetyl group is the acyl group of acetic acid. Its chemical formula is CH_3CO, sometimes written as Ac (not to be confused with the chemical element actinium Ac). Remember that an acyl or an acyl group is a radical or functional group obtained by removing the hydroxyl group OH from a carboxylic acid.

Second stage

Deprotection of the hydroxyl functions (transition from acetyl groups AcO to hydroxyl groups OH) can be carried out either in an acid medium (HCl) or a basic medium (NaOH). The reaction equation is written as:

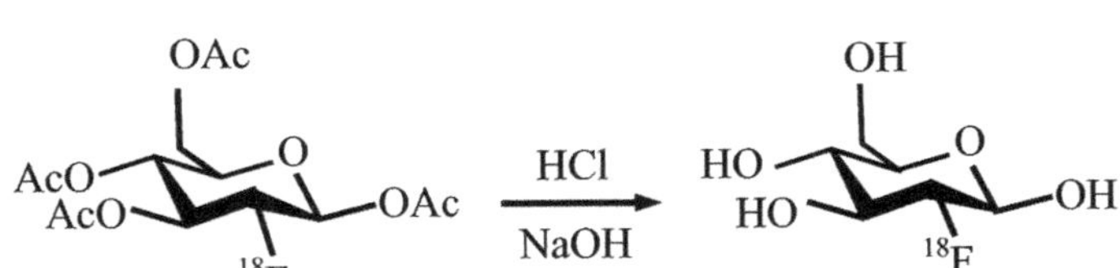

[3.9]

2-Deoxy-2-[^{18}F]fluoro-D-glucose
(^{18}FDG)

2-stage yield: 60–65%

NOTE.–

The (Tf) group of 1,3,4,6-tetra-*O*-acetyl-2-*O*trifluoromethane sufonyl-β-D-mannopyranose 1 (Figure 3.4) is explained by writing this formula in the following possible forms [VAL 09, JOY 13, AND 18]:

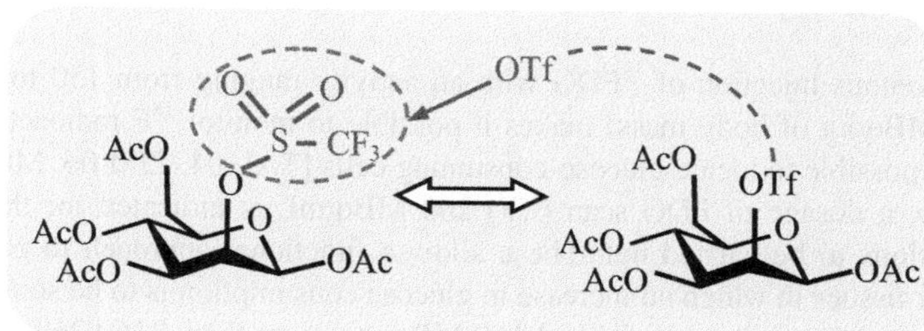

Figure 3.4. *Two possible formulas for 1,3,4,6-tetra-O-acetyl-2-O-trifluoromethane sufonyl-β-D-mannopyranose. For a color version of this figure, see www.iste.co.uk/sakho/nuclearphysics3.zip*

Finally, let us note the few security measures adopted during the preparation of ^{18}F and ^{18}FDG. We refer to the above-mentioned preparations using the equipment used by the Groupement d'Intérêt Public, GIP CYROI (Cyclotron Réunion Océan Indien) that was established in 2004 [GRO 15].

The cyclotron for the manufacture of fluorine-18 is placed in a containment vessel with walls and ceiling made of 2 m thick concrete. Regarding the pharmaceutical manufacturing unit, the incorporation of the radioactive fluorine atom into the sugar molecule is automated in chambers shielded by 80 mm of lead. The production unit has a quality control laboratory to check the purity of the drug before it is injected into humans. As far as radiation protection is concerned, taking into account the short half-life of fluorine-18, the decay of solid and liquid waste is the preferred solution for its disposal. A very small amount of material can be contaminated with radioelements with a much longer half-life, most often resulting from the activation of certain parts of the cyclotron. These are isolated and stored in a decay room, then evacuated and treated. The gases that result from the synthesis are collected in airtight bags (balloons). After decaying for at least 24 h, these gases are evacuated into the extraction circuit, whose filters trap the particles present [GRO 15].

3.1.4. *Principle of PET Scan, dosimetry*

As shown in equation [3.9], ^{18}FDG is a sugar in which one of the hydroxyl groups (alcohol function) has been substituted by a fluorine-18 atom. The injectable solution is a clear, colorless to pale yellow liquid. Consider, for example, a volume of 1 mL of injectable solution contains 260 MBq - (^{18}F) at the date and time of calibration. The total activity per vial is between 260 MBq and 4,160 MBq; the usual recommended activity for adults is 100 to 400 MBq (depending on the patient's body mass and the type of camera used) administered by direct intravenous injection [AGE 09].

Intravenous injection of ^{18}FDG with an activity ranging from 150 to 550 MBq (2 to 6 MBq/kg of body mass) makes it possible to monitor ^{18}F radioactivity. This makes it possible to locate glucose-consuming cells [TAL 04, ZIM 06, MON 18]. In oncology, a dosage of FDG scan (^{18}F) 260 MBq/mL is indicated for the imaging examinations to be carried out. These allow a functional approach to pathologies, organs or tissues in which an increase in glucose consumption is to be sought. As the maximum total activity per vial is 4,160 MBq, no more than 4,160/260 = 16 mL of FDG scan (^{18}F) 260 MBq/mL should be administered. It should be noted that in addition to the designation FDG scan (^{18}F), the term fluorodeoxyglucose positron emission tomography (PET-scan ^{18}F-FDG) is used in the literature. The acronym PET stands for positron emission tomography.

Table 3.2 [TOU 01] shows the dosimetry used for ^{18}FDG scintigraphy, according to the ICRP (*International commission on radiological protection*). It should be noted that the dose received by the patient depends on their age and the organ being scanned.

^{18}F	**2- Fluoro-2-deoxy-D-glucose (FDG)**				
	Intravenous injection				
	Absorbed dose per unit of activity administered				
	(µGy/MBq)				
Organs	**Adult**	**15 years**	**10 years**	**5 years**	**1 year**
Bladder wall	160	210	280	320	590
Myocardium	62	81	120	200	350
Brain	28	28	30	34	82
Ovaries	15	20	30	44	82
Testes	12	16	26	38	73
Uterus	21	26	39	55	100
Effective dose (µSv/MBq)	**19**	**25**	**37**	**51**	**95**

Table 3.2. *Absorbed doses in ^{18}FDG scintigraphy. For a color version of this table, see www.iste.co.uk/sakho/nuclearphysics3.zip*

It should be noted that the patient must be prepared for the injection session at least the day before the scintigraphy examination. The following measures must be respected [AGE 09]:

1) FDG scan (^{18}F) 260 MBq/mL must be stored and handled in suitable shielded enclosures in order to provide maximum protection for patients and hospital staff. In particular, it is advisable to protect against the effects of β^+ and annihilation photon irradiation by using a syringe shield when distributing the contents of the vial and packaging the syringe.

2) FDG scan (^{18}F) 260 MBq/mL should be administered to a well-hydrated patient who has been fasting for at least 4 h to obtain maximum uptake of the radioactive glucose analogue, as intracellular glucose transport is limited by a "saturable mechanism". Water intake should not be restricted (no sugary drinks).

3) In order to obtain good-quality images and reduce bladder irradiation, patients should be advised to drink copiously and empty their bladder before the images are acquired, and frequently after the examination.

4) In order to avoid hyperfixation of the tracer in the muscles, it is recommended to avoid any significant physical activity before the examination and to remain at strict rest between the injection and the examination and during image acquisition (patient lying comfortably, without reading or talking).

5) Lastly, blood sugar levels should be checked prior to administration, as hyperglycemia, especially when it is greater than 8 mmol/L, may reduce the sensitivity of the FDG scan (^{18}F) 260 MBq/mL PET scan. For the same reason, administration of this product should be avoided in patients with unbalanced diabetes.

In nuclear medicine imaging, ^{18}F-FDG is used as a tracer in positron emission tomography (PET). It is primarily metabolized by cells that consume a lot of glucose, particularly in the brain and liver, as well as by cancer cells. The general principle of scintigraphy using ^{18}F-FDG is shown in Figure 3.5.

Monitoring ^{18}F radioactivity makes it possible to locate glucose-consuming cells. Regions in which glucose metabolism is higher than expected will thus be highlighted, with the area of tracer activity being in the lungs [POI 19].

Fluorine-18 has an activity of 3.52×10^{18} Bq per gram, emitting positrons as shown in Figure 1.6. The particles β^+ of 633.5 keV then encounter electrons in the material through which they pass. The annihilation of an electron–positron pair then produces two γ photons of the same energy 511 keV (Figure 3.6).

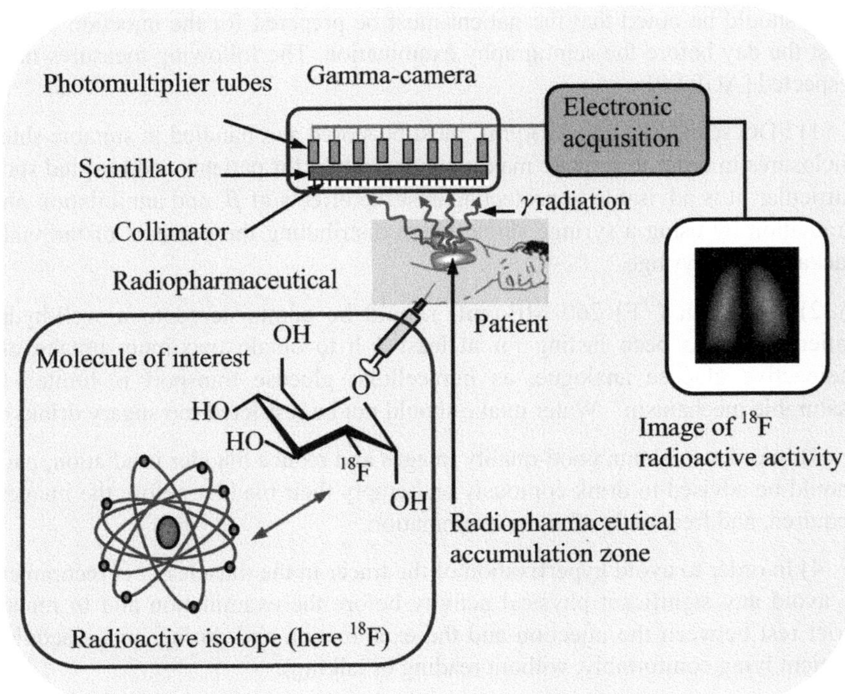

Figure 3.5. *Schematic diagram of an ^{18}FDG scintigraphy. For a color version of this figure, see www.iste.co.uk/sakho/nuclearphysics3.zip*

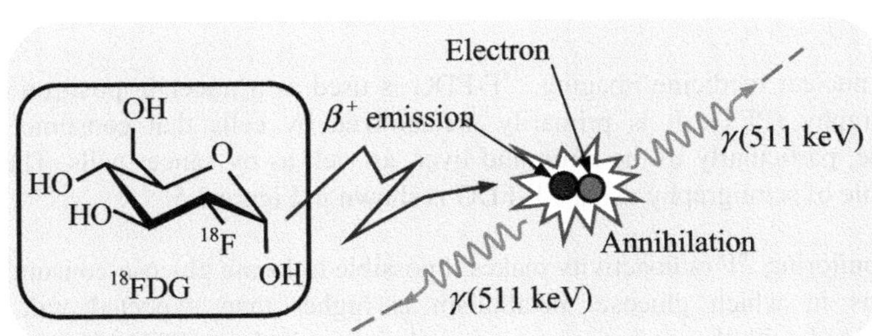

Figure 3.6. *Production of two photons γ following the annihilation of a positron emitted by ^{18}FDG via ^{18}F radiofluorine and an electron from the material. For a color version of this figure, see www.iste.co.uk/sakho/nuclearphysics3.zip*

As shown in Figure 3.6, once the radiopharmaceutical is attached to its target, it emits γ radiation, which is detected by a *gamma-camera* (*scintillation camera*). The gamma camera (Figure 3.5) consists of three components: a *collimator*, a *scintillator* and *photomultiplier tubes*. This assembly is linked to acquisition electronics and a computer that analyzes all the data collected and displays the 2D or 3D image of the targeted part of the body.

In practice, image acquisition usually begins 45–60 min after the injection of fludeoxyglucose-(^{18}F). Provided that activity in the tissues (tumor, heart, brain) has reached a plateau and that sufficient activity is still present to obtain an adequate count rate, image acquisition can also be carried out 2–3 h after administration, which reduces background noise [AGE 09].

The SPECT (*single-photon emission computed tomography*) imaging used in *nuclear cardiac stress* is carried out using gamma cameras. Usually one, two or three detectors or heads rotate slowly around the patient's torso. The first gamma camera was developed in 1957 by Hal Anger, and can measure the radioactivity of an entire surface at once. Anger's gamma cameras are based on the principle of indirect conversion: the incident gamma photon is converted into light photons and then into an electrical signal. This method uses a scintillating sodium iodide (NaI) crystal and photomultiplier tubes [BLA 17].

NOTE.–

Some PET cameras offer the option to choose between 2D and 3D acquisition modes. 2D mode performs significantly better than 3D mode in overall image quality for patients with a body mass index (BMI) > 27.5 [ISN 07].

Note that the body mass index, more commonly known as BMI, is calculated from height and weight, according to the following formula: BMI = weight in kg/height^2 (in m). This is used to assess weight status. According to the WHO, a normal BMI should be between 18.5 and 24.9 (see: https://www.livi.fr/sante/imc/).

BIOGRAPHY CORNER.–

Hal Oscar Anger (1920–2005) was an American electrical engineer and biophysicist. He is best known for inventing the scintillation camera, known as the gamma camera (γ-camera) or Anger camera, in 1957. He also developed the well counter, widely used in laboratory tests to measure radioactivity in samples. Anger also developed a multi-plane tomographic radiation scanner using the Anger camera and a focused radiation collimator.

3.1.5. *Excretion of fluorine-18 from the body*

When ingested by the cell, ^{18}FDG undergoes phosphorylation (metabolic transformation), preventing it from leaving the tumor cell. This leads to an accumulation of all the identical molecules that have been assimilated. The result is a kind of amplification of the radioactive signal, which becomes increasingly intense. The part of the ^{18}FDG not consumed by the cells is very rapidly eliminated via the urine. The fluorine trapped in the cells decays rapidly. Following the rapid decay of this isotope, around 20 h after the injection, well under one thousandth of the initial radioactive dose is still detectable in the body [ZIM 06].

It should be noted that the International Commission on Radiological Protection (ICRP) indicates that in FDG form, fluorine-18 is distributed in the heart (4%), the brain (8%) and the liver (5%), where retention is permanent, with the remaining 80% in other organs and tissues. 25% of the injected activity is excreted via the urine, with biological periods of 12 min (25%) and 1.5 h (75%) [INS 13a].

NOTE.– Administration of ^{18}FDG to pregnant or breastfeeding women.

Unless otherwise indicated, there is no clinical experience to-date on the use of ^{18}FDG in pregnant women. However, when a radiopharmaceutical is to be administered to a fertile patient (i.e. of childbearing age), it is advisable to enquire about the possibility of pregnancy, considering any woman who has not menstruated until proven otherwise to be pregnant. As a precautionary measure, examinations (CT scans, ultrasound scans, etc.) that do not require the use of radioactive tracers should be preferred, as imaging techniques using radiopharmaceuticals in pregnant women also involve irradiation of the fetus.

It is acknowledged that the administration of (^{18}F) 260 MBq/mL at an activity of 400 MBq results in an absorbed dose of 8.4 mGy in the uterus. At these doses, no lethal effects, induction of malformations, growth delays or functional disturbances are expected. However, the risk of inducing cancer or hereditary deficiencies could be increased. An FDG scan (^{18}F) 260 MBq/mL should therefore not be administered during pregnancy unless PET is absolutely necessary or when the benefit to the mother outweighs the risk to the fetus [AGE 09].

In the case of breast-feeding women, FDG scan (^{18}F) 260 MBq/mL is excreted in breast milk following injection. Therefore, before administering ^{18}FDG to a woman who is breast-feeding, it is wiser to consider postponing the scintigraphy examination until the end of breast-feeding (otherwise use a bottle in an extreme emergency). However, if it is essential to administer the scan while breastfeeding, then breastfeeding should be suspended for at least 12 h after administration of the FDG scan (^{18}F) 260 MBq/mL. Thereafter, the breast must be emptied to eliminate the milk produced during the 12 h after administration (it may be possible to pump the milk before injecting the ^{18}FDG and keep it to feed the baby during and after the examination period).

APPLICATION 3.3.–

An oncology hospital is looking to analyze glucose metabolism in lung cancer. The radiopharmaceutical used is ^{18}FDG. A patient is injected with 185 MBq of ^{18}FDG. There is a time delay of two and a half hours between the industrial preparation of ^{18}FDG and its use in the hospital center. Determine the activity of the radiopharmaceutical at the time of its industrial preparation.

ANSWER.–

A_0 activity is equal to:

$$A_0 = A(t)e^{\ln 2\ t/T}. \qquad [3.10]$$

At the time of injection (2h30 min = 150 min previously), the activity of radiopharmaceutical A = 185 MBq. For ^{18}F, the half-life T = 109.8 min (Figure 1.1). The activity at the time of industrial preparation is equal to:

$$A_0 = 185e^{\ln 2\ (150/109.8)} = 476.886 \text{ MBq} \approx 477 \text{ MBq}.$$

APPLICATION 3.4.–

Consider the radiopharmaceutical ^{18}FDG with an activity of 185 MBq to be injected into a patient. What is the corresponding radiopharmaceutical mass?

Data:

– *Avogadro number*: $N_A = 6.02 \times 10^{23}$ mol^{-1};

– *molar mass of ^{18}FDG*: $M = 181$ g · mol^{-1}.

ANSWER.–

Let us use the relations defining the mol number n to express the number N of nuclei:

$$n = \frac{m}{M} = \frac{N}{N_A} \Rightarrow m = \frac{N}{N_A}M. \qquad [3.11]$$

Knowing that $A = \lambda N$, we obtain:

$$m = \frac{AT}{\ln 2}\frac{M}{N_A}. \qquad [3.12]$$

NUMERICAL APPLICATION.–

$$m = \frac{185 \times 10^6 \times 109.8 \times 60}{\ln 2}\frac{181}{6.02 \times 10^{23}} \approx 5.3 \times 10^{-10}\ \text{g}$$

3.2. Properties of the ^{81m}Kr radiotracer, dosimetry

Lung scintigraphy is a non-invasive, painless medical imaging test that is indicated in cases of breathlessness and uses low radioactive doses [LAC 11, CAR 17, REI 18, DEL 19]. This scintigraphy is a dual examination which enables the essential functions of the lungs, namely perfusion (blood circulation) and ventilation, to be explored in detail.

During the first phase of ventilation, the patient inhales a radiotracer, which is a gas or aerosol containing a weakly radioactive product. In the case of ventilation scintigraphy, metastable krypton-81 (^{81m}Kr) or aerosols labeled with metastable technetium-99 (^{99m}Tc) are used as the radioactive gas [CIM 79, BOM 12, BLA 18]. The properties of metastable technetium-99 were studied in section 2.2. In what follows, we are interested in studying the properties of metastable krypton-81.

3.2.1. *Production of the ^{81m}Kr tracer by the ^{81}Rb–^{81m}Kr generator*

Krypton-81m is produced using an ^{81}Rb–^{81m}Kr generator. Irradiation is carried out on natural krypton (4.5% Kr) with protons of about 30 MeV using a remotely controlled gas target [CIM 79]. Two nuclear transmutation reactions are used to obtain ^{81}Rb: ^{82}Kr $(p, 2n)$ ^{81}Rb and ^{83}Kr $(p, 3n)$ ^{81}Rb. The first, dominant reaction is explicitly written as:

$${}_{1}^{1}H + {}_{36}^{82}Kr \rightarrow 2\,{}_{0}^{1}n + {}_{37}^{81}Rb \qquad [3.13]$$

The main contaminant is rubidfum-82m, with a half-life of 6.4 h. This decays 6% by positron emission β^+ and 94% by electron capture. Its emission spectrum consists of several lines with energies between 0.511 and 1.475 MeV [CIM 79].

The principle of the $^{81}Rb/^{81m}Kr$ generator is based on the fixation of Rb on a column of (cationic) ion exchange resin. Impurities are removed by washing the column. Rubidium-81 (^{81}Rb), with a half-life of 4.58 h, is absorbed on filter paper in the form of a rubidium chloride solution. It decays to form gaseous krypton-81m (^{81m}Kr) with a particularly short half-life of 13 s:

$$^{81}_{37}Rb \rightarrow \; ^{81m}_{36}Kr + \, ^{0}_{1}e + \nu_e \qquad [3.14]$$

The krypton-81m formed then decays by isomeric transition into krypton-81, emitting pure gamma radiation of 190 keV according to the equation:

$$^{81m}_{36}Kr \rightarrow \; ^{81}_{36}Kr + \gamma \qquad [3.15]$$

The krypton-81 generated by the reaction [3.15] decays by electron capture with a half-life of 2.1×10^5 years to give stable bromine-81 (^{81}Br) [THO 18]. Note that the ^{81m} Kr formed by decay of ^{81}Rb can also be eluted either by moist air (ventilation generator) or by an isotopic glucose solution (perfusion generator) [CIM 79].

3.2.2. *^{81m}Kr tracer ventilation scintigraphy: dosimetry*

The ^{81m}Kr obtained via reaction [3.14] has a number of advantages. Its very short half-life of 13 s minimizes the risk of undesirable irradiation of the patient. Used without a vector, ^{81m}Kr can be used to perform a ventilation test, thus complementing the perfusion test performed with human albumin macroaggregates (MAA). Furthermore, krypton is chemically inert (it is a rare gas), so there is no danger to the patient. In nuclear medicine imaging, ^{81m}Kr emits a single γ radiation of 190 keV, thus ensuring excellent detection and very good images with a NaI detector [THO 18]. Table 3.3 [TOU 01] shows the dosimetry used in the case of ^{81m}Kr scintigraphy introduced into the body by inhalation.

In ventilation scintigraphy, the ^{81m}Kr administration system used consists of six components:

1) a flexible tube;

2) a three-way breathing valve;

3) an antibacterial filter;

4) a mouthpiece or breathing mask;

5) a large-section discharge tube;

6) a decay container.

^{81m}Kr	**Krypton Inhalation Absorbed dose per unit of activity administered (μGy/MBq)**				
Organs	**Adult**	**15 years**	**10 years**	**5 years**	**1 year**
Lungs	0.21	0.31	0.44	0.68	1.3
Breasts	< 0.01	< 0.01	< 0.01	0.013	0.018
Pancreas	< 0.01	< 0.01	< 0.01	< 0.01	0.018
Ovaries	< 0.01	< 0.01	< 0.01	< 0.01	< 0.01
Testes	< 0.01	< 0.01	< 0.01	< 0.01	< 0.01
Uterus	< 0.01	< 0.01	< 0.01	< 0.01	< 0.01
Effective dose (μSv/MBq)	**0.03**	**0.04**	**0.06**	**0.09**	**0.16**

Table 3.3. *Absorbed doses in ^{81m}Kr ventilation scintigraphy. For a color version of this table, see www.iste.co.uk/sakho/nuclearphysics3.zip*

The flexible tube consists of a central tube with a small cross-section and an outer tube with a larger cross-section. The central tube allows the eluted krypton to pass through. The outer tube supplies the air to be inhaled and acts as an ^{81m}Kr reservoir during exhalation.

In addition, the flexible tube connected to a generator is linked to a three-way breathing valve. The remaining two ports of the valve are connected to the mouthpiece (or breathing mask) via an antibacterial filter, and to the outer tube, evacuating the effluent to the decay container. This container, fitted with appropriate protective shielding, is used to collect the exhaled radioactive gas until complete decay of the ^{81m}Kr with a half-life of 13 s.

The instructions for use of the ^{81m}Kr administration system are as follows [AGE 11]:

– connect the flexible tube to the generator using the connections;

– connect the power plug to 220 V (a light indicates correct connection);

– connect the other end of the flexible tube to the inlet of the three-way valve and connect it, on the patient side, to the antibacterial filter (the inlet, outlet and patient side of the valve are each clearly indicated by arrows on the valve);

– connect the large cross-section tube for the exhaled air to the outlet of the three-way valve and to the decay container;

– put a nose clip on the patient when using the mouthpiece;

– connect the mouthpiece or breathing mask to the antibacterial filter;

– allow the patient to breathe deeply and normally. Ensure that the patient completes all breathing cycles via the mouthpiece or mask;

– switch on the pump and start image acquisition. The air supply pump for the krypton-81m elution generator can be switched on or off using the switch on the control panel or the remote control;

– stop the pump immediately after taking the images in order to stop the release of krypton-81m.

It should be noted that of the various tracers used in ventilation scintigraphy, krypton-81m has a clear advantage over its competitors in terms of patient dosimetry and radiation protection of personnel. In addition, Kr-81m is a gas, so bronchial deposits are not visible on the scintigraphic images, as can be the case with Technegas, and especially with [^{99m}Tc]-DTPA [LAC 11].

3.3. Lung scintigraphy examination

Unlike stress myocardial scintigraphy, where it is preferable for the patient to fast the day before the stress test, in the case of lung scintigraphy, no special precautions need to be taken by the patient. Patients can eat and drink before and after the test and take their medication as usual. There is also no need to undress. However, any metal objects (keys, jewelry) that could affect image quality should be removed [LAC 11, CAR 17, REI 18, DEL 19].

3.3.1. *Procedure for the ventilation test*

The following is a summary of the main information on lung ventilation scintigraphy:

– *Utility*: precise exploration of the essential functions of the lungs (ventilation test followed by a perfusion test).

– *Radiopharmaceutical used*: low-level radioactive aerosol labeled with Technetium-99m or Krypton-81m.

– *Patient preparation*: no fasting required; patient can lie still during the examination.

– *Examination procedure*: breathing in the radioactive aerosol: 15 min; ventilation image under the gamma camera: 15 min, i.e. 30 min for the entire examination.

Now let us look at the practicalities of the test.

The patient breathes in a low-level radioactive aerosol labeled with Technetium-99m or Krypton-81m for approximately 15 min. During this time, the camera records an initial series of images (the properties of aerosols are discussed in section 3.5).

Before the examination, the patient lies on their back. They then inhale a product containing an isotope (radioactive gas labeled with Krypton-81m or aerosol labeled with technetium-99m), which is weakly radioactive and has a very short duration of action (Figure 3.7) [BLA 18].

Figure 3.7. *Patient inhaling a product containing a weakly radioactive isotope during a pulmonary ventilation scintigraphy examination. For a color version of this figure, see www.iste.co.uk/sakho/nuclearphysics3.zip*

The inhaled radiotracer follows the permeable airways from the mouth to the alveoli, as shown in Figure 3.8.

While the patient breathes normally and effortlessly, inhaling the radioactive product administered, the gamma camera revolves around the patient, recording images for a few minutes without touching them. If one or more areas are not properly ventilated, the product is not delivered, resulting in images in which the tracer is not present and which indicate obstructed ventilation. The clinical data associated with the images will point to a tumor or inflammation such as *chronic obstructive pulmonary disease* (COPD) [CAR 17, REI 18, BLA 18].

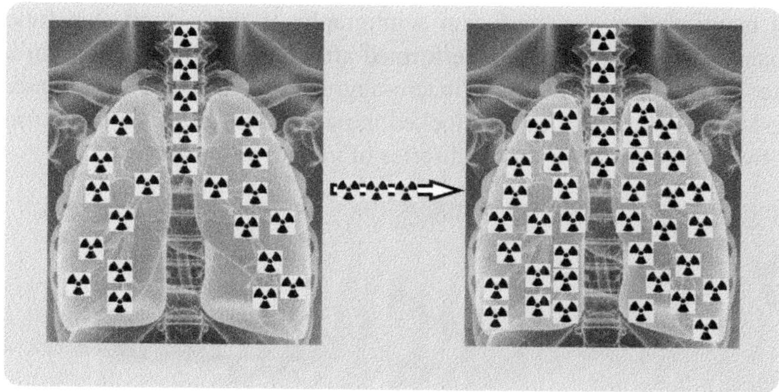

Figure 3.8. *Path of the radiotracer inhaled by the patient along the permeable airways, from the mouth to the alveoli. For a color version of this figure, see www.iste.co.uk/sakho/nuclearphysics3.zip*

VOCABULARY CORNER.–

Chronic obstructive pulmonary disease, or *COPD* for short, is a group of chronic systemic diseases of respiratory origin, affecting the bronchi.

A *bronchus* is a duct that brings oxygen-rich air from outside the body into the lungs.

3.3.2. *Perfusion examination procedure*

The main information on lung perfusion scintigraphy is given below:

– *Utility*: precise exploration of the essential functions of the lungs (ventilation test followed by a perfusion test).

– *Radiopharmaceutical used*: ^{99m}Tc-labeled human albumin macroaggregates (MAA).

– *Patient preparation*: no fasting required; patient must lie still during the examination.

– *Test procedure*: IV injection of the radiopharmaceutical, perfusion image under the gamma camera: 10 min.

Now let us look at the practicalities of the test.

As mentioned above, perfusion scintigraphy is the second step following the pulmonary ventilation test. It is performed after the injection into a vein in the arm (Figure 3.9) of human albumin macro-aggregates labeled with Technetium 99m (^{99m}Tc-MAA). The particles are blocked in part of the *arteriolar capillaries* of each of the two lungs, enabling a second series of images to be taken.

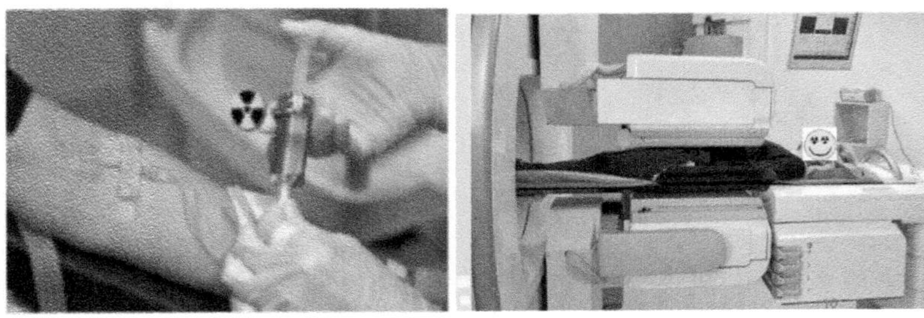

Figure 3.9. *Patient undergoing scintigraphy examination [BLA 18]. Injection of ^{99m}Tc-MAA into a vein in the arm, then passage under the gamma camera moving around the patient, while recording the images for a few minutes (approximately 10 min). For a color version of this figure, see www.iste.co.uk/sakho/nuclearphysics3.zip*

VOCABULARY CORNER.–

Capillaries: capillaries are the finest and smallest blood vessels in vertebrates. Their role is to *supply tissues with nutrients and oxygen*. Connected to veins and arteries, blood capillaries transport nutrients, particularly glucose, and oxygen. They are very present in the muscles, liver, lungs, kidneys and nervous system [GIO 21].

Arteriole: an arteriole is a blood vessel whose diameter is less than 40 or 100 μm (micrometer). Its diameter depends on the degree of arteriolar contraction resulting from the branching of the arteries and driving blood towards the capillaries.

3.3.3. *Elimination of ^{99m}Tc and ^{81m}Kr in the body*

In normal pulmonary circulation, MAA particles are distributed throughout the lungs following a physiological gradient. The time spent by ^{99m}Tc-MAA in the lungs varies according to the structure, size and number of particles injected.

In the lungs, MAA particles decrease according to an exponential decay law as a result of ^{99m}Tc disintegration. The products of ^{99m}Tc-MAA disintegration, returned to the circulation in the form of albumin microcolloids, are rapidly eliminated by macrophages of the reticuloendothelial system, mainly in the liver and spleen. Microcolloids are metabolized, releasing the radioactive tracer (^{99m}Tc) into the bloodstream, where it is eliminated and excreted in the urine [AGE 13b]. Similarly, the ^{81m}Kr tracer is cleared via the urine. It is therefore advisable to drink plenty of water to facilitate *elimination* of the radiopharmaceutical used during scintillation.

3.4. Radiopharmaceuticals used in thyroid scintigraphy

3.4.1. *Benefits of iodine*

Iodine is an essential chemical element in thyroid hormones. Hence, it is easily fixed by the thyroid gland. In the body, iodine normally comes from food and water, so normal dietary intake makes it difficult to ingest excessive quantities of iodine. The iodine absorbed into the body via the diet is present in the form of I^- iodide ions, which are concentrated in the follicular cells from the blood using the Na/I symport. This support uses the electrochemical gradient in sodium ions Na^+ to accumulate I^- ions, which are used to produce thyroid hormones.

The particular benefit of radiotherapy with iodine is linked to the fact that only thyroid cells can absorb it (other organs do not store it). Iodine not stored in the thyroid is eliminated in a short time by the kidneys and from there in the urine. Due to these particular properties of iodine circulation in the body, a very high radiation dose can be obtained in the target tissue, while the rest of the body is exposed to only a small amount. Parathyroid scintigraphy is an examination that studies the functioning of the parathyroid glands, using radioactive products injected in small quantities. It is used to study the origin of certain parathyroid function disorders (adenoma, hyperplasia. etc.). The parathyroid glands are located behind the thyroid gland [RIM 21]. Thyroid function is diagnosed in nuclear medicine by injection of radioactive iodine-123.

VOCABULARY CORNER.–

Hormone: substance produced by a gland. Hormones influence the development or functioning of an organ. They are a way for our bodies to communicate information throughout the body, essentially by means of molecules carried by the blood. There are a large number of hormones, which are essential for the body to function properly. Each hormone has a specific function, such as stimulating growth, managing stress, controlling fertility, regulating body temperature, etc. Sometimes, they stimulate the growth of cancer cells.

3.4.2. ^{123}I radiotracer production chain

As shown in Table 4.4 in Chapter 4 of Volume 2, the radioisotope ^{123}I, with a half-life of 3.18 days, is produced either directly or indirectly in a cyclotron. Several nuclear transmutation reactions are used to produce radioiodine-123. These include [FUS 72, LAM 72, KAH 77, INS 09b, ALA 16, HER 19]:

– $^{127}\mathrm{I}\ (p, 5n)\ ^{123}\mathrm{Xe} \rightarrow {}^{123}\mathrm{I}$;

– $^{121}\mathrm{Sb}\ (\alpha, 2n)\ ^{123}\mathrm{I}$; $^{122}\mathrm{Te}\ (d, n)\ ^{123}\mathrm{I}$; $^{122}\mathrm{Te}\ (\alpha, 3n)\ ^{123}\mathrm{Xe} \rightarrow {}^{123}\mathrm{I}$.

The production reactions for ^{123}I are explained below.

– *Indirect reaction $^{127}I\ (p, 5n)\ ^{123}Xe \rightarrow {}^{123}I$*

Proton bombardment of iodine-127 produces xenon and five neutrons. The latter decays to iodine 123 by 87% electron capture and 13% β^+ emission. The iodine-123 formed decays to 100% by electron capture to give excited tellurium-123. The latter de-excites to its fundamental level by emission of a 159 keV photon. The equations for the reactions leading to the formation of iodine-123 and stable tellurium-123 are given below.

– *Transmutation of iodine-127*

$$^{1}_{1}H + {}^{127}_{53}I \rightarrow 5\,{}^{1}_{0}n + {}^{123}_{54}Xe \qquad [3.16]$$

– *Disintegration of xenon-123 (by electron capture and β^+ emission)*

$$^{123}_{54}Xe + {}^{0}_{-1}e \rightarrow {}^{123}_{53}I + X + \nu_e : 87\% \qquad [3.17a]$$

$$^{123}_{54}Xe \rightarrow {}^{123}_{53}I + {}^{0}_{1}e + \nu_e : 13\% \qquad [3.17b]$$

– *Disintegration of iodine-123 and de-excitation of tellurium-123*

$$^{123}_{53}I + {}^{0}_{-1}e \rightarrow {}^{123}_{52}Te* + X + \nu_e : 100\% \qquad [3.18a]$$

$$^{123}_{52}Te* \rightarrow {}^{123}_{52}Te + \gamma \qquad [3.18a]$$

The various nuclear reactions [3.16], [3.17] and [3.18] are summarized in Figure 3.10.

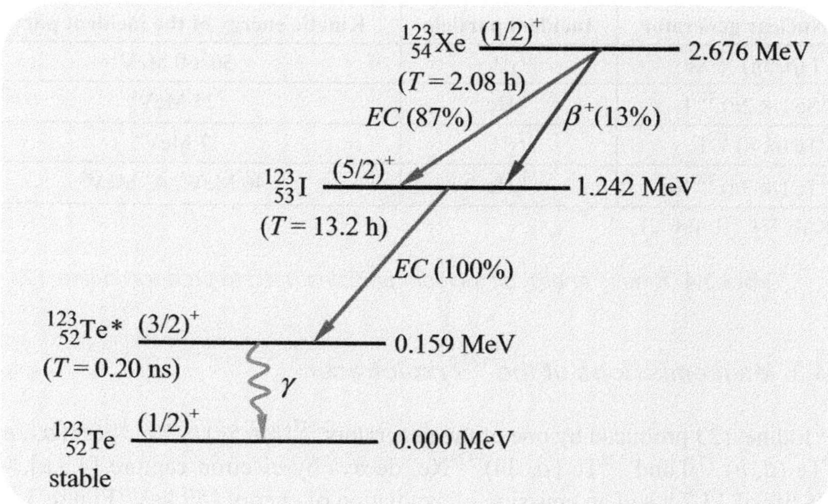

Figure 3.10. *Formation of iodine-123 via the nuclear reaction ^{127}I (p, 5n) ^{123}Xe. For a color version of this figure, see www.iste.co.uk/sakho/nuclearphysics3.zip*

– *Direct reactions ^{121}Sb (α, 2n) ^{123}I and ^{122}Te (d, n) ^{123}I*

By bombardment of antimony-121 by a flux of α particles (^{4}He) or of tellurium-122 by a flux of deuterons (^{2}H), iodine-123 is obtained according to the reactions:

$$^4_2He + ^{121}_{51}Sb \rightarrow 2\,^1_0n + ^{123}_{53}I \quad [3.19a]$$

$$^1_1H + ^{122}_{52}Te \rightarrow ^1_0n + ^{123}_{53}I \quad [3.19b]$$

– *Indirect reaction ^{122}Te (α, 3n) $^{123}Xe \rightarrow ^{123}I$*

– *Transmutation of tellurium-122*

$$^4_2He + ^{122}_{52}Te \rightarrow 3\,^1_0n + ^{123}_{54}Xe \quad [3.19c]$$

– *Disintegration of xenon-123 according to equations [4.4.c]*

Table 3.4 summarizes some of the characteristics of the particles involved in the direct (or indirect) production of iodine-123.

Nuclear generator	Incident particle	Kinetic energy of the incident particle
$^{127}I\ (p, 5n)\ ^{123}Xe$	$^{1}_{1}H$	50–60 MeV[a]
$^{121}Sb\ (\alpha, 2n)\ ^{123}I$	$^{4}_{2}He$	25 MeV[a]
$^{122}Te\ (d, n)\ ^{123}I$	$^{2}_{1}H$	7 MeV[a]
$^{122}Te\ (\alpha, 3n)\ ^{123}Xe$	$^{4}_{2}He$	46 MeV[a]; 42 MeV[b]

[a] [KAH 77]; [b] [LAM 72]

Table 3.4. *Kinetic energy of incident particles used to produce iodine-123*

3.4.3. Main emissions of the ^{123}I radiotracer

Iodine-123 produced by one of the generators $^{127}I\ (p, 5n)\ ^{123}Xe$, $^{121}Sb\ (\alpha, 2n)\ ^{123}I$, $^{122}Te\ (d, n)\ ^{123}I$ and $^{122}Te\ (\alpha, 3n)\ ^{123}Xe$, decays by electron capture [4.3.a], with a half-life of 13.2 h and an emission of γ radiation of energy 159 keV (Figure 3.10).

As explained in Chapter 4 of Volume 1 in section 4.1.7 [SAK 22], the electron capture (EC) phenomenon competes with β^+ decay. This phenomenon is accompanied by the emission of X photons as a result of electronic rearrangement. In addition, the emission of X photons generally competes with the emission of Auger electrons through a process known as the Auger effect. In the case of iodine-123, the main emissions observed with percentages greater than 1% [INS 09b] are shown in Table 3.5.

Emitted particles	Energy (in keV)	Emission percentage (%)
X photon	3.8	3.4
	4.0	2.1
	27.2	25.0
	27.5	46.0
	30.9	13.2
	31.7	2.4
Electron	127.0	14.0
	154.0	2.0
γ photon	159.0	83.0
	529.0	1.4

Table 3.5. *Main emissions of iodine-123 in percentages greater than 1%*

Note the emission of Auger electrons at an energy of between 2 and 5 keV (98%) and between 21 and 32 keV (12.4%) [INS 09b],

3.4.4. *Properties of [^{123}I] MIBG and [^{123}I] ioflupane radiopharmaceuticals*

In nuclear medicine, iodine-123 is mainly used for in vivo diagnosis in SPECT imaging (in particular, thyroid scintigraphy) in the functional study of the thyroid gland. It acts as a tracer in the most widely used radiopharmaceuticals. These are ^{123}I-metaiodobenzylguanidine (^{123}I-MIBG) ADREVIEW ® and ^{123}I-ioflupane Datscan® [INS 01]. These radiotracers have the respective crude formulas $C_8H_{10}IN_3$ and $C_{18}H_{23}FINO_2$. The corresponding semi-developed formulas are I – C_6H_4 – CH_2 – NH – CNH(NH_2) and F – $(CH_2)_3$ – N – C_7H_{10} – CO(OCH_3) – C_6H_4I. In the literature, their structural formulas are presented as shown in Figure 3.11.

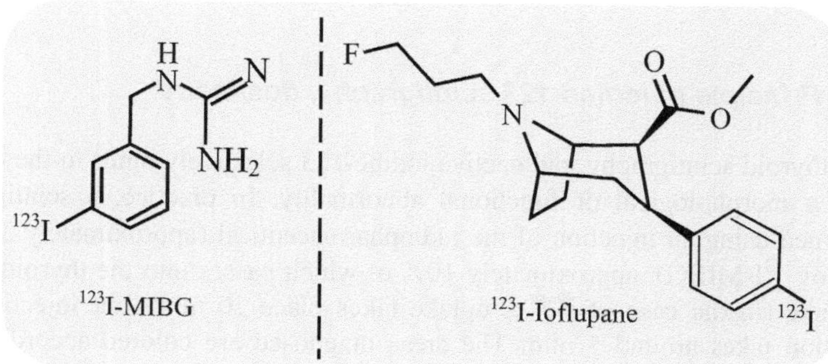

Figure 3.11. *Structures of the radiopharmaceuticals ^{123}I-metaiodobenzylguanidine (^{123}I-MIBG) with the formula $C_8H_{10}IN_3$ and ^{123}I-ioflupane with the formula $C_{18}H_{23}FINO_2$*

In practice, ^{123}I-MIBG is used in children to diagnose neuroblastoma, while ^{123}I-ioflupane is used in adults to diagnose parkinsonian syndrome.

Radioiodine-123 has the same chemical properties as natural iodine, which is captured by the thyroid gland. In the specific case of thyroid scintigraphy, examinations based on the use of iodine-123 are the most widespread in nuclear medicine because the energy of the emitted γ radiation equal to 159 keV (Figure 3.10 and Table 4.2) is very suitable for detection using a gamma camera centered on the patient's thyroid. This gamma camera is equipped with a pinhole collimator to obtain a magnification effect (due to the very small size of the thyroid gland).

VOCABULARY CORNER.–

Neuroblastoma: cancer of the nervous system that mainly affects young children and most often develops in the abdomen. Neuroblastoma is a cancer that develops from immature nerve cells called neuroblasts. During the normal process of differentiation, neuroblasts give rise to neurons of the autonomic nervous system (ANS), part of the nervous system which triggers the body's automatic actions (breathing, digestion, heart function, sweating – profuse sweating). In the case of neuroblastoma, this maturation process malfunctions and the neuroblasts multiply uncontrollably and then accumulate to form a tumor (see: https://www.fondation-arc.org/cancer/neuroblastomes/quest-ce-quun-neuroblastome, 2023).

Parkinsonian syndrome: this syndrome refers to symptoms of Parkinson's disease. It is a brain disease that causes motor problems, mental health and sleep disorders, etc. (for more details on this subject, see Volume 2, Appendix 3).

3.4.5. *Principle of iodine-123 scintigraphy, dosimetry*

In thyroid scintigraphy, radioactive iodine-123 selectively binds to the thyroid to detect a morphological or functional abnormality. In practice, a scintigraphy is performed using an injection of the radiopharmaceutical (approximately 3.7 MBq $\cdot$ kg^{-1} of ^{123}I-MIBG), approximately 10% of which passes into the thyroid after 2 h of uptake (in the case of ^{99m}Tc, uptake takes place 20 min after injection). Data collection takes around 5 min. The areas diagnosed are colored according to the degree of fixation of the radioactive marker used. It should be noted that the only disadvantage of iodine-123 is its high production cost, as generators must be used via cyclotrons.

Table 3.6 [TOU 01] shows the dosimetry used for ^{123}I-MIBG scintigraphy.

The radiopharmaceutical ^{123}I-ioflupane (Datscan®) is used in practice for the differential diagnosis of essential tremor. Datscan has long been used for the differential diagnosis between *essential tremor* and degenerative parkinsonian syndrome (origin of the tremor) and between Alzheimer's disease and Lewy body disease (origin of the cognitive disorders). By extension, it has become a second-line diagnostic tool for the diagnosis of Parkinson's disease by exclusion of the other signs: if there is no essential tremor, no Alzheimer's disease and no Lewy body disease, then there is a high probability of having *idiopathic Parkinson's disease* in the presence of *striatal markers* [PER 14, THO 18]. Parkinson's disease, Alzheimer's disease and dementia with Lewy bodies are discussed in the appendices to Volume 2.

^{123}I	**Metaiodobenzylguanidine (MIBG) [1] Intravenous injection Absorbed dose per unit of activity administered (µGy/MBq)**				
Organs	**Adult**	**15 years**	**10 years**	**5 years**	**1 year**
Liver	67	87	130	180	330
Bladder wall	48	61	78	84	150
Spleen	20	28	43	66	120
Ovaries	8,2	11	16	25	46
Testes	5,7	7,5	12	18	33
Uterus	10	13	20	29	53
Effective dose (µSv/MBq)	**13**	**17**	**26**	**36**	**67**

[1] The estimated absorbed dose does not take into account the presence of radiocontaminants. Given the radiochemical purity of the radionuclides currently available, the influence of radiocontaminants on the effective dose is minimal,

Table 3.6. *Absorbed doses in [^{123}I] metaiodobenzylguanidine (^{123}I-MIBG) scintigraphy. For a color version of this table, see www.iste.co.uk/sakho/nuclearphysics3.zip*

Essential tremor is a neurological disease that differs from neurodegenerative diseases such as Parkinson's, Alzheimer's and dementia with Lewy bodies. In fact, no region of the brain loses its neurons. It therefore involves involuntary, rhythmic, reciprocal oscillatory movements of antagonistic muscle groups, generally involving the hands, head, face, vocal cords, trunk or legs. Diagnosis is clinical [HEC 22, LAM 20]. Essential tremor is an action tremor, both postural (e.g. when a limb is held in a fixed position against gravity) and intentional (e.g. when a movement is initiated).

Furthermore, essential tremor is a disease that progresses slowly over several years, can worsen over time and can be very disabling in daily activities. Essential tremor affects both men and women, and can appear in early childhood or early old age. When it appears in childhood, it progresses throughout life, peaking in adolescence. It is common between the ages of 55 and 65. Essential tremor often progresses slowly and cannot be stopped once it has been diagnosed.

Half of all cases are hereditary (hence the importance of asking patients about their family history). Essential tremor currently affects 1 in 200 of the world's population. In France, more than 300,000 people suffer from these tremors, three to four times more than sufferers of Parkinson's disease, and 30,000 of them have a severe, even disabling form [LAM 20].

VOCABULARY CORNER.–

– *Stratum*: the stratum is the inner part of the brain that regulates motivation and impulses, and is the main target of dopamine. It is probably the most important area of the brain in decision-making and also plays a key role in addiction.

– *Addiction*: pathology based on the repeated consumption of a product (tobacco, alcohol, drugs, etc.) or the abnormally excessive practice of a behavior (games, time spent on social networks, etc.), which leads to:

- a loss of control over the level of consumption/practice;
- a change in emotional balance;
- medical problems;
- disruption to personal, professional and social life.

– *Striatal markers*: radiotracers used to explore the stratum. Example: fluorodopa (^{18}F), or ^{18}FDOPA (Figure 3.12), is a fluorinated form of L-DOPA labeled with fluorine-18 and used as a radiotracer. It is administered by intravenous injection and is used in PET scans to visualize dopaminergic nerve endings in the striatum for the evaluation of suspected Parkinson's syndrome.

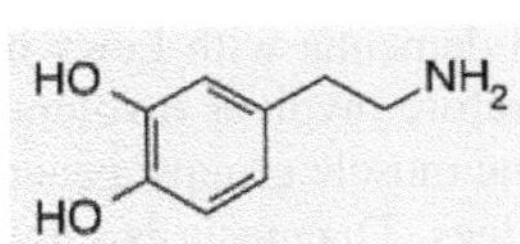

Dopamine

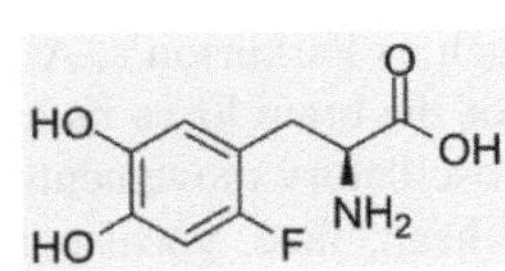

L-6-[^{18}F]fluoro-DOPA

Figure 3.12. *Structures of dopamine and L-6-[^{18}F]fluoro-DOPA*

Dopaminergic pathways, sometimes called dopaminergic projections, are the set of projection fibers in the brain that synthesize and release the neurotransmitter (chemical compound released by neurons) dopamine. The individual neurons in these pathways are called dopaminergic neurons.

Dopamine is a neurotransmitter that enables communication within the nervous system, and plays a direct role in influencing behavior. Dopamine reinforces normally beneficial actions, such as eating a healthy food, by provoking a sensation of pleasure. It is essential for individual survival.

– *Idiopathic Parkinson's disease (IPD)*: accounts for the vast majority of Parkinson's disorders. In fact, this form alone affects 75% of sufferers. This is the true form of Parkinson's disease. This form of the disease can be distinguished in part from other Parkinsonian symptoms, but its main difference lies in its unidentified causes.

3.4.6. *Excretion of iodine-123 from the body*

If the body receives iodine supplements, 25% of this intake will, on average, be taken up directly by the thyroid. This percentage is closely linked to the normal daily intake from the diet. When the daily intake is reduced, the amount taken up can easily reach 50%; whereas, with a high daily intake, it can fall to 5–10%. The remainder is eliminated by the body fairly rapidly, within a few days, mainly in the urine, as well as in other excretions such as feces, sweat, saliva and respiration [DIR 99]. However, [^{123}I-MIBG is eliminated mainly by the urinary route, 30% of the quantity of product fixed in the thyroid gland is eliminated over a period of 13 h. Between 40% and 50% of ^{123}I-MIBG is excreted 24 h after injection, and 70% to 90% is recovered in the urine after 96 h [JIM 19]. It should be noted that the ^{123}I-MIBG fraction distributed outside the thyroid (70%) is mainly eliminated by the kidneys, a small proportion being eliminated in the feces, saliva, sweat and exhaled air [INS 09b].

NOTE.–

Most free iodine is concentrated in the thyroid gland. Concentrations peak around 24 h after intake. In pregnant women, iodine is also transferred to the fetus, where it accumulates in the thyroid gland. In breast-feeding women, it accumulates in the milk, mainly in the form of iodide, via the mammary gland. Labeled iodine follows the metabolism of the vector to which it is attached. Due to its lipophilic nature, it is very well absorbed through the skin [INS 09b].

APPLICATION 3.5.–

A thyroid scan is performed using a sodium iodide NaI solution containing the 123I isotope, The nuclear medicine center orders a vial of NaI solution. Injections are made every 30 min, with the first injection taking place at 9 a.m. Just before this injection, the activity in the vial was equal to 28.5 MBq. At each injection, a dose of 7 MBq is taken. What is the residual activity of the solution after the second injection? It will be assumed that the activity of the solution in the vial does not vary at the time of the first injection.

ANSWER.–

From Table 3.4 in Chapter 3, we derive the period of radioiodine-123: $T = 67.2$ h. In addition, the dose taken during the first injection has an activity of 7 MBq. The initial activity of the solution just before the second injection is $A_0 = (28.5 - 7)$ MBq $= 21.5$ MBq. The second injection takes place 30 min = 0.5 h later. The residual activity of the solution in the vial just before the second injection is then equal to:

$$A = A_0 e^{-\ln 2\, t/T} = 21.5\, e^{-\ln 2\,(0.5/67.2)} = 21.1 \text{ MBq}. \qquad [3.20]$$

APPLICATION 3.6.–

The previous scintigraphic examination is reconsidered. Determine the number of patients who can receive a dose of 7 MBq.

ANSWER.–

Let n be the number of patients. The total injectable dose verifies the inequality (regardless of the 30 min between injections):

$$28.5 - 7n > 6 \Rightarrow n < (28.5 - 6)/7 = 3.21.$$

A maximum of three patients underwent thyroid scintigraphy.

We can also reason as follows:

Let us assume that all of the injections are given simultaneously at a dose of 7 MBq per patient. Since the initial activity is equal to 28.5 MBq, a total of 4 patients can be administered. This corresponds to 4×7MBq $= 28$ MBq. This leaves 0.5 MBq. Activity decreases over time. Let us find the residual activity at the time of the fourth injection, i.e. one and a half hours after the first injection. This gives us:

– first IV: $A_0 = (28.5 - 7)$ MBq $= 21.5$ MBq;

– second IV: $A_0 = 21.1$ MBq [3.20]; we take $(21.1 - 7) = 14.1$ MBq for the second IV;

– third IV: $A = 14.1\, e^{-\ln 2\,(0.5/67.2)} = 14.0$ MBq; we take $(14.0 - 7) = 7$ MBq for the third IV;

– fourth IV: $A = 7\, e^{-\ln 2\,(0.5/67.2)} = 6.96$ MBq < 7 MBq (this result must not be rounded): a dose of 7 MBq cannot be taken.

3.4.7. ^{131}I *radiotracer production chain*

Like iodine-123, iodine-131 is also used in nuclear medicine. In the treatment of differentiated thyroid cancers [HER 19], total removal of the gland is generally followed by treatment with iodine-131 to eradicate the thyroid remnants left during surgery [FER 11]. For this reason, it is important to study its properties and uses in nuclear imaging. We will also study the properties of iodine-125 and its use in the specific case of prostate brachytherapy using iodine-125 implants (see section 3.6). In scintigraphy, the ^{131}I radiotracer is injected in the form of sodium iodide Na^{131}I for the treatment of thyroid cancer [HER 19].

Iodine-131 can be obtained by fission of uranium-235 (see note, equation [4.5]) or by neutron bombardment of tellurium-130 in a nuclear reactor [INS 09a, FER 11, AUS 11, RUI 13, HER 19]. In practice, a TeO_2 tellurium oxide target enriched with tellurium-130 is used [AUS 11]. By bombarding the tellurium-130 contained in this target with neutrons, two nuclear reactions ^{130}Te (n, γ) ^{131m}Te and ^{130}Te (n, γ) ^{131g}Te are carried out, written in the condensed form ^{130}Te (n, γ) ^{131m,g}Te. The reaction equations are written as:

$$ {}_{0}^{1}n + {}_{52}^{130}Te \rightarrow \gamma + {}_{52}^{131m}Te \qquad [3.21a] $$

$$ {}_{0}^{1}n + {}_{52}^{130}Te \rightarrow \gamma + {}_{52}^{131g}Te \qquad [3.21b] $$

Metastable tellurium-131 (^{131m}Te), with a half-life of 30 h, and ground-state tellurium-131 (^{131g}Te), with a half-life of 25 min, decay in the β^- mode to give iodine-131. The corresponding equations are as follows:

$$ {}_{52}^{131m}Te \rightarrow {}_{53}^{131}I + {}_{-1}^{0}e + \bar{\nu}_e \qquad [3.22a] $$

$$ {}_{52}^{131g}Te \rightarrow {}_{53}^{131}I + {}_{-1}^{0}e + \bar{\nu}_e \qquad [3.22b] $$

It should be noted that in the reactor, 22% of the ^{131m}Te formed is transformed into ^{131g}Te [FER 11].

NOTE.–

In the work by Ruivo et al. [RUI 13], a mass of 5–20 mg of natural tellurium was irradiated by a thermal neutron flux equal to 3.2×10^{12} n/cm^{2}s, for a few minutes to generate the ^{130}Te(n, γ) ^{131}Te reaction in the IEA-R1 nuclear reactor at the IPEN/CNEN-SP center in São Paulo, Brazil [FRA 09, RUI 13]. In their experimental studies, Ruivo et al. found half-lives equal to 33.25 $\pm$ 0,25 h for ^{131m}Te and 18.89 ± 0,09 min for ^{131g}Te, compared with the generally accepted values of 30 h and 25 min [FER 11, AUS 11, HER 19].

3.4.8. *Main emissions of iodine-131*

Iodine-131 is 100% β^- radioactive with a half-life of 8.0207 (1) d [QAI 11]. Its decay yields xenon-131, which de-excites with a maximum emission of about 82% γ-photons of about 364 keV. The decay also produces electrons emitted at up to 606 keV with a percentage of around 90%. The decay equation is written as:

$$^{131}_{53}I \rightarrow {}^{131}_{54}Xe + {}^{0}_{-1}e + \bar{\nu}_e + \gamma \qquad [3.23]$$

The decay diagram of iodine-131 is shown in Figure 3.13.

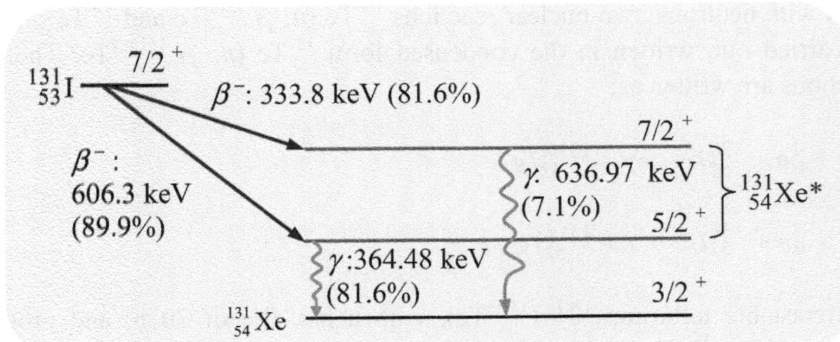

Figure 3.13. *Decay diagram of iodine-131. For a color version of this figure, see www.iste.co.uk/sakho/nuclearphysics3.zip*

In Figure 3.13, we have indicated the β^- transitions corresponding to the maximum decay energies of 333.8 keV and 606.3 keV. In addition, the de-excitation processes of the $^{131}Xe^*$ son nucleus corresponding to the most energetic photons (636.97 keV and 364.48 keV) are shown. Table 3.7 shows the main emissions [INS 09a] observed in the decay process of iodine-131.

NOTE.–

As we said earlier, iodine-131 can be obtained by fission of uranium-235 in nuclear reactors. During a nuclear accident, ^{131}I can be released into the environment via the reaction:

$$^{1}_{0}n + {}^{235}_{92}U \rightarrow {}^{131}_{53}I + {}^{99}_{39}Y + 6\,{}^{1}_{0}n \qquad [3.24]$$

Emitted particles	Energy (in keV)	Emission percentage (%)
γ photon	80.18	2.6
	284.30	6.2
	364.48	81.6
	636.97	7.1
	722.89	1.8
Electron	45.6	3.5
	329.9	1.5
Beta (E_{max})	247.9	2.1
	333.8	7.2
	606.3	89.9
X photon	29.4	1.5
	29.7	2.8

Table 3.7. *Main emissions of iodine-131*

Radioactive iodine-131, produced by the fission of uranium, can enter the bloodstream by inhalation, through the skin or by eating contaminated food. In the body, the thyroid gland accumulates either iodine-131 or ordinary iodine-127 to saturation. A considerable proportion of the iodine-131 intake will be concentrated in the thyroid within 24–48 h. During the initial retention and recycling period when radioactive iodine is in the body, it irradiates thyroid tissue, resulting in the death of tumor cells in the case of *thyroid cancer* [DIR 99].

Prolonged decay of iodine-131 in the gland increases the risk of thyroid cancer (see section 3.8.2), with fetuses, babies and young children running the greatest risk. During nuclear accidents that release iodine-131 into the environment, people exposed to polluted air can take ordinary iodine tablets to prevent the body from accumulating radioactive iodine-131. This is because when the thyroid gland is saturated with ordinary iodine, it does not accumulate radioactive iodine. Excess iodine in the body due to the presence of iodine-131 will simply be eliminated via the urine and feces.

3.4.9. *Principle of iodine-131 scintigraphy: dosimetry*

Iodine-131 is mainly used for medical applications. However, it can be used as an environmental radiotracer after a nuclear accident (see note). In radiotherapy, iodine-131 in the form of sodium iodide $Na^{131}I$ is indicated for the treatment of thyroid cancer [HER 19]. A capsule containing iodine-131 in the form of $Na^{131}I$ has an activity varying between 37 and 7400 MBq at the date and time of calibration [AGE 13b]. The capsule is administered orally on an empty stomach and should be swallowed whole with plenty to drink. For pediatric populations, the capsule may be crushed with food to ensure that it is swallowed whole, when administered.

In the treatment of thyroid cancer or metastases, hyperthyroidism and goiter, the desired effect is the destruction of all or part of the cells of the thyroid gland or metastases. Damage to other organs should not produce deterministic effects [DIR 99]. As shown in Table 3.7, iodine-131 releases a β^- particle with a maximum energy of 606.3 keV and the highest emission percentage of 89.9%. It is this β^- radiation with a mean path of 0.5 mm in the tissue that is responsible for the therapeutic activity.

In addition, γ radiation with a maximum energy of 364.48 keV (81.6% of the photons emitted) emitted by de-excitation of the $^{131}Xe^*$ son nucleus (Figure 3.13) can leave the thyroid gland and induce undesirable irradiation effects on the patient and their environment. However, it has the advantage of being able to be used to control the fixation of radioactive iodine in the thyroid gland. It should be noted that the β^- particles emitted during the disintegration of iodine-131 are ionizing particles that cause damage to the DNA molecule located in the immediate vicinity of the thyroid gland, in particular, double strand breaks, leading to the programmed death of the cell. We already know that in 2000, Professor Léon Sanche's team at the University of Sherbrooke in Canada demonstrated that low-energy electrons (LEEs) or sub-ionization electrons (0–15 eV) are capable of breaking bonds in the DNA molecule via resonant dissociative attachment processes. Our current research also focuses on the study of the effect of low-energy electrons (LEE) on the DNA molecule for applications in radiobiology [SAK 23a], and on the assessment of the risks of radiation-induced cancers by the out-of-field dose in external radiotherapy [NGU 23].

In practice, the activity of iodine-131 to be administered is assessed by the clinician and depends on the cause of the hyperthyroidism, the size of the thyroid gland, thyroid uptake and iodine clearance. Table 3.8 [TOU 01] shows the dosimetry used for scintigraphy with iodine-131 in the form of sodium iodide $Na^{131}I$.

^{131}I	**Sodium iodide** **Intravenous injection** [1] **Absorbed dose per unit of activity administered (µGy/MBq)**				
	Thyroid fixation: 25%				
Organs	**Adult**	**15 years**	**10 years**	**5 years**	**1 year**
Thyroid	360.000	560.000	840.000	1.900.000	3.400.000
Stomach wall	460	580	850	1.500	3.000
Bladder wall	460	570	870	1.400	2.600
Small intestine wall	280	350	320	1.000	2.000
Ovaries	43	58	91	140	260
Testes	27	34	56	92	180
Uterus	52	70	110	170	300
Effective dose (µSv/MBq)	**18.122**	**28.158**	**42.238**	**95.394**	**170.744**
	Thyroid fixation: 0%				
Organs	**Adult**	**15 years**	**10 years**	**5 years**	**1 year**
Bladder wall	610	750	1.100	1.800	3.400
Kidneys	65	80	120	170	310
Colon wall	40	48	75	124	219
Stomach wall	34	40	64	100	190
Ovaries	42	54	84	130	240
Testes	37	45	75	120	230
Uterus	54	67	110	170	300
Effective dose (µSv/MBq)	**64**	**78**	**120**	**193**	**365**

[1] For oral administration, the dose to the stomach wall is increased by approximately 30%. The effective dose is not significantly altered

Table 3.8. *Doses absorbed in [^{123}I] metaiodobenzylguanidine (^{123}I-MIBG) scintigraphy. For a color version of this table, see www.iste.co.uk/sakho/nuclearphysics3.zip*

NOTE.–

Sodium iodide [^{131}I] is contraindicated in the event of pregnancy, suspicion of pregnancy or when pregnancy has not been ruled out (the radiation dose absorbed in the uterus for this product is probably between 11 and 511 mGy; in addition, the fetal thyroid is particularly avid for iodine during the second and third trimesters). When it is necessary to administer radiopharmaceuticals to a woman of childbearing age, any possibility of pregnancy must be ruled out (any woman who has not had her period must be considered pregnant until proven otherwise). Women receiving sodium iodide [I-131] should be advised not to become pregnant within 6 to 12 months of administration. Women are also advised to use contraception for a period of 6 to 12 months. As a precaution, men should not father a child for 6 months after treatment with radioactive iodine, to allow irradiated spermatozoa to be replaced by non-irradiated spermatozoa. Before administering a radiopharmaceutical to a woman wishing to continue breast-feeding, it is necessary to assess whether this treatment can be deferred until the end of breast-feeding. After administration of sodium iodide [^{131}I], breastfeeding should be stopped permanently [AGE 13b]. As a precautionary measure, it should be remembered that imaging techniques using radiopharmaceuticals in pregnant women also involve irradiation of the fetus.

3.4.10. *Excretion of iodine-131 from the body*

As mentioned above, the thyroid gland does not differentiate between ordinary iodine-127 and radioactive iodine-131. The main daughter product of radioactive iodine-131 is xenon-131 [3.23], which is rapidly eliminated from the body. In the body, half of the quantity of iodine-131 decays in 8 days. Damaged thyroid cells lose their ability to bind iodine, so it is returned to the bloodstream, leading to excretion. Radioactive iodine is therefore eliminated from the body fairly rapidly, either by radioactive decay or by metabolic excretion. The total quantity of radioactive iodine is reduced to half its initial value at a rate that depends on the state of the disease: from 1 day for thyroid cancer with total removal of thyroid tissue, to 7 days for patients with a benign goiter.

In the case of hyperthyroidism, the effective period is about 4–5 days [DIR 99]. In Figure 3.14, we have reproduced the iodine-131 excretion curves for thyrotoxicosis, cancer treatment and cancer follow-up. This figure shows the differences in the rate of excretion and retention in the body (which is linked to it) over the 20 days following administration [DIR 99].

For the eradication of hyperthyroidism, the doses administered to patients must be relatively high, known as "ablative", in the order of 370 to 600 MBq (10 to 16.2 mCi) or 5.9 to 6.5 MBq (160 to 176 μCi) retained per gram of thyroid tissue [TOU 18]. Depending on the patient's general condition and the degree of thyrotoxicosis, medical preparation with synthetic antithyroid drugs (ATS) may be proposed. However, it is preferable to stop taking ATS medication 4–7 days before the cure so as not to limit its efficacy [TOU 18, NWA 19]. In the work of Nwatsock et al. [NWA 19], the therapeutic activity of iodine-131 was adapted to thyroid volume. Patients with a thyroid volume less than or equal to 25 grams received 740 MBq, those with a volume between 25 and 40 grams received a fixed activity of 1,110 MBq, and when the thyroid volume was greater than 40 grams, an activity of 1,480 MBq was administered.

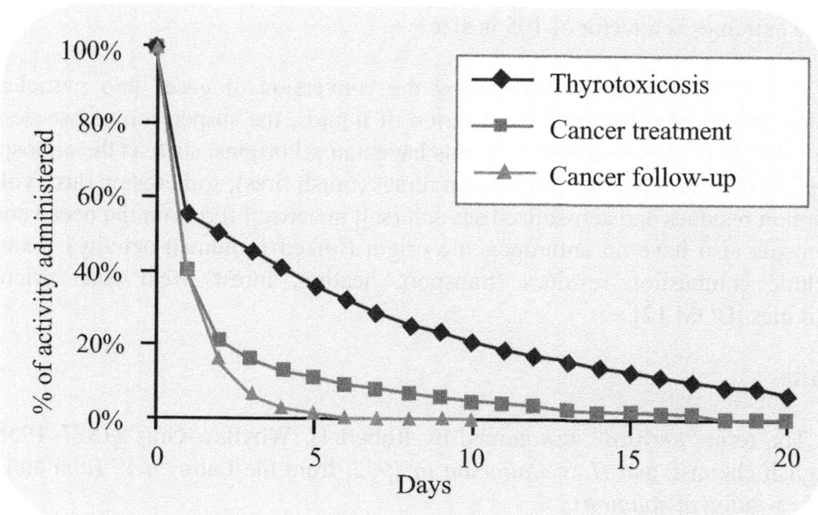

Figure 3.14. *Iodine-131 excretion curves as percentages of radioactivity administered for thyrotoxicosis, cancer treatment and cancer follow-up (metastases or recurrence). For a color version of this figure, see www.iste.co.uk/sakho/nuclearphysics3.zip*

As a reminder, incidence is the number of new patients over a given period, while prevalence measures the proportion of the population affected by the disease at a given time.

3.5. General information on aerosols

As seen, aerosols labeled with metastable technetium-99 (^{99m}Tc), such as Technegas ® labeled with ^{99m}Tc and the metastable krypton-81 (see Chapter 2, section 2.4), are used in lung scintigraphy. For this reason, it is important to know the properties of aerosols and how they are inhaled by mouth or nose.

3.5.1. *Aerosol therapy: the concept of aerosol*

Aerosol therapy is a medical procedure involving the delivery of an active medicinal ingredient by inhalation using aerosols. An aerosol is a suspension of solid or liquid particles in a gas, forming a two-phase system. The size range is between ~1 nm (the size of large molecules) and 100 µm (beyond which the particles do not remain in suspension and settle rapidly). The difference between these extremes is a factor of 105 in size.

Aerosols are naturally formed by the conversion of gases into particles, the disintegration of solids, the nebulization of liquids, the suspension of powders and the splitting of agglomerates. Aerosols have natural origins, such as the atmospheric droplets that form clouds, combustion fumes (brush fires), soil erosion dust, volcanic eruption residues and aerosolized sea salt (salt in aerosol form) on the ocean surface. Aerosols also have an anthropogenic origin (linked to human activity). Examples include combustion residues (transport, heating, forest fires) and agricultural activities [BOM 12].

NOTE.–

The term "*aerosol*" was coined by Robert H. Whytlaw-Gray (1877–1958), an English chemist, and *H. S. Patterson in 1932,* from the Latin "*aer*" (*air*) and "*sol*" (abbreviation of *solution*).

3.5.2. *Nebulization system: aerosol generators*

Nebulization is a process used to administer radioactive substances to a patient by inhalation, directly to the target tissues. The target tissues are located in the bronchial wall, the ENT (*ear, nose and throat*) area or the deep lung. Briefly, the *nebulization system consists of* an aerosol generator with a nebulization tank, a delivery circuit consisting of a ringed tube located between the generator and the patient interface, and possibly ancillary heating, timer, alarm and other systems. *In general, there are two types of aerosol dispenser*: *pneumatic* and *ultrasonic*. The choice of generator depends on the pathology to be treated and the nature of the radioactive product to be administered.

The pneumatic generator consists of a nebulizer and a pressure source. The ultrasonic generator is a device designed to emit ultrasonic waves (sound waves with frequencies $f > 20$ kHz propagate in the air with the same speed as sound equal to $340 \text{ m} \cdot \text{s}^{-1}$). These waves are produced by the oscillations of quartz subjected to the effect of an electric current in a liquid. These vibrations cause fine liquid particles to detach from the surface. This then creates a mist that is sent to the patient via ventilation. The size of the particles generated is fixed and depends on the quartz vibration frequency.

It is worth noting that ultrasonic devices generate a higher percentage of particles smaller than 5 µm than pneumatic devices (around 70% for the former, compared with 50–60% for the most efficient of the latter), thus enabling deeper penetration of the respiratory tract [BON 10].

3.5.3. *Notion of median mass aerodynamic diameter (MMAD)*

Aerosols, particularly those used for medical applications, are characterized by a quantity defined by European standard EN 13544-1, the median mass aerodynamic diameter (MMAD). This corresponds to the particle size value for which 50% by mass of the particles in the aerosol are larger and 50% are smaller. The deposition and impaction of particles along the airway will depend directly on this MMAD. Particles with an MMAD greater than 5 µm will be mainly deposited in the upper aerodigestive tract by impaction, those with an MMAD between 2 and 6 µm will reach the tracheobronchial level by sedimentation, and particles between 0.5 and 3 µm will reach the pulmonary alveoli by diffusion. Very fine particles (< 0.5 µm) will be exhaled. One of the particularities of using a radioactive aerosol is that, as the quantity of radioactivity carried by each aerosol particle is proportional to its volume, which increases by a factor of 3 with the size of the particles, there will be much greater activity in the larger particles than in the smaller ones. To avoid the visualization of hot spots corresponding to the impaction and accumulation of radioactive particles along the respiratory tract during ventilation scintigraphy, it is necessary to use nebulization systems. These allow the generation of the finest particles possible, and ideally the MMAD should be less than 1.2 µm [LAC 11].

3.5.4. *Particle deposition phenomena*

In the general case, the deposition of aerosols is observed in cylindrical ducts. Three main phenomena can then be observed: impaction, sedimentation and diffusion, as shown in Figure 3.15 [BEC 10]. The characteristics of these three phenomena are as follows [NER 09, BON 10, BEC 10]:

– *Impaction* occurs with aerosols of MMAD > 5 µm. In ducts, impaction is due to a deviation of the particles from the fluid flow lines under the effect of the centrifugal force acting on each of the particles. This inertial deviation is due to the presence of a bend in the duct and the presence of a wall on the new particle trajectories beyond the bend.

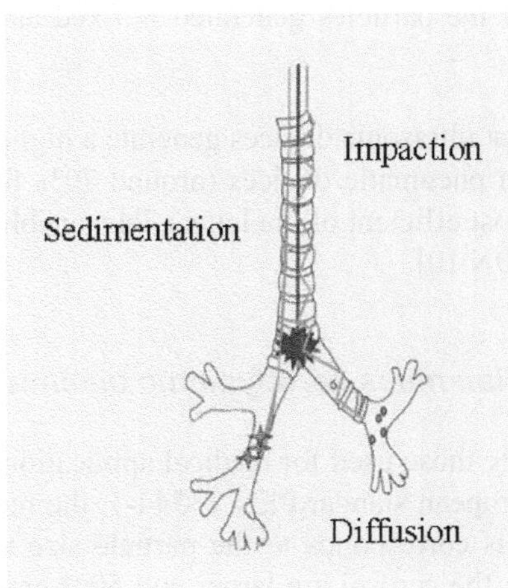

Figure 3.15. *Illustration of aerosol deposition mechanisms by impaction, sedimentation and diffusion. For a color version of this figure, see www.iste.co.uk/sakho/nuclearphysics3.zip*

In addition, the phenomenon of fluid turbulence can also lead to the deposition of aerosols by impaction, a deposition mechanism equally inertial in nature. Turbulence is explained by the fact that a given fluid particle that has received sufficient energy from a granular environment with a turbulent structure can then be projected, thanks to this energy input, until it meets a wall to deposit or, in other words, to impact.

– *Sedimentation* occurs with aerosols whose MMAD is between 2 and 5 μm. The phenomenon of particle drift due to sedimentation is explained by the combined action of the force of gravity and the force of drag acting on each of the particles in the fluid, causing them to settle on the wall. The force of gravity is the resultant of the gravitational force (weight of the particle) and Archimedes' buoyancy, which opposes the weight of the volume of fluid displaced. This resultant is sometimes called the buoyancy force. The drag force is the major force exerted on a particle in a fluid. The drag force opposes the relative movement between the particle and the fluid that surrounds it, and thus acts as a fluid friction force.

– *Diffusion* occurs for particles with an MMAD of between 1 and 3 μm. This diffusion can be interpreted as a Brownian effect diffusion mechanism. In fact, in a fluid, the agitation of particles in all directions (random agitation) gives rise to a force called the Brownian force. This force results from the effect of shocks between a fluid particle and the other particles it encounters in its environment. Note that the influence of the Brownian force increases as the MMAD of the particles decreases.

3.5.5. *Aerosol deposition sites and methods*

The location of aerosol deposition also depends on the MMAD of the particles making up the aerosol and the inspiratory flow rate. It also depends on the phenomenon of aerosol deposition [BEC 10, POI 20].

Drug deposition also depends on the speed at which the particles move "in situ". Flow is rapid in the upper airways, decreasing progressively as we move towards the small bronchi and pulmonary alveoli. Particles are deposited by impaction when the velocity is high in the nasopharynx (*upper pharynx, behind the nose*), where the heaviest particles are deposited.

As speed decreases, deposition is more likely to occur by sedimentation, i.e. by gravity, within the small bronchi and will involve medium-sized droplets. However, the smallest droplets will diffuse into the alveoli by being thrown against the walls due to the Brownian movements of the gas (molecular agitation).

Flow is rapid in the upper airways, decreasing progressively as we move towards the small bronchi and pulmonary alveoli [BON 10].

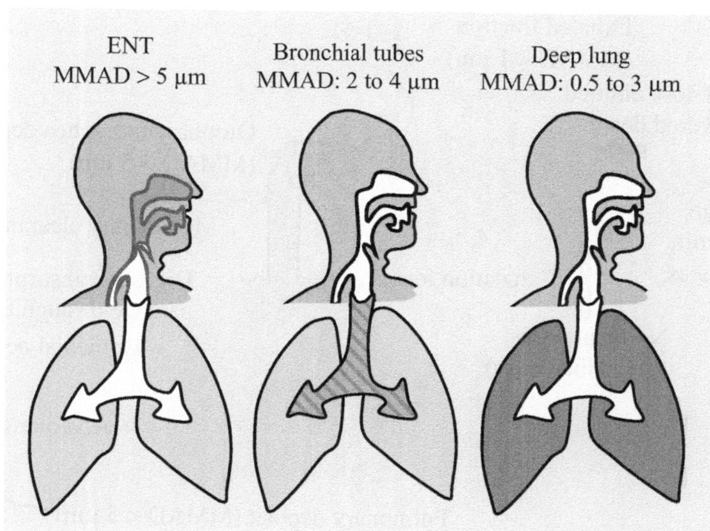

Figure 3.16. *Deposition of inhaled aerosol particles according to their MMAD. For a color version of this figure, see www.iste.co.uk/sakho/nuclearphysics3.zip*

After inhalation (rapid inhalation through the nose or mouth), the larger particles with an MMAD > 5 μm (under the effect of centrifugal force) will be deposited by impaction in the ENT area and the large bronchi. Following slow, deep inhalation, fine particles with an MMAD of between 1 and 5 μm (due to the combined effects of gravity and drag) will be deposited by sedimentation, mainly in the bronchial regions. Finally, with slow, deep inspiration through the mouth followed by *apnea* (momentary suspension of breathing), particles with an MMAD of between 0.5 and 3 μm (diffused by the Brownian effect) will be deposited mainly in the deep lung (Figure 3.16).

It should be noted that flow rate also influences where aerosols are deposited. A high inspiratory flow rate or bronchial obstruction promotes the proximal deposition of large particles; whereas, slow, deep inspiration promotes the distal deposition of medium and small particles.

A large airflow volume and apnea at the end of inspiration are conducive to the deposition of fine particles [BON 10]. In addition, sitting or standing while administering an aerosol promotes sedimentation of aerosols towards the base of the airways to the detriment of the apices [BEC 10].

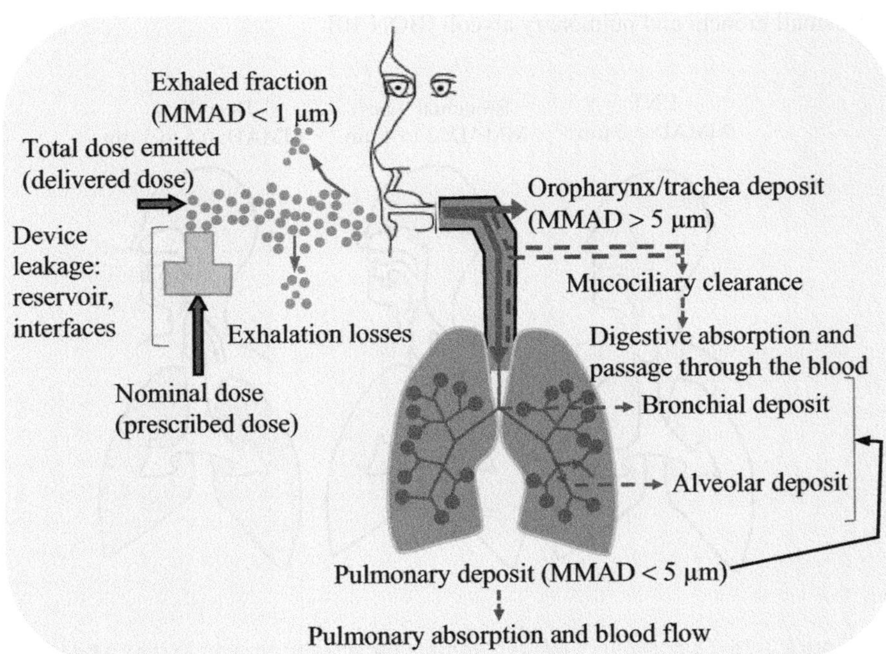

Figure 3.17. *Drug pathway during the nebulization session. For a color version of this figure, see www.iste.co.uk/sakho/nuclearphysics3.zip*

Figure 3.17 summarizes the path taken by aerosols through the various deposition sites until they reach the bloodstream. In addition to *mucociliary clearance*, aerosol particles are eliminated from the body by the kidneys after passing into the blood.

VOCABULARY CORNER.–

– *Mucociliary system*: the body's defense mechanism, which comes into play each time we breathe in. As the air is not sterile, the *kinocilia* in the nose prevent certain micromolecules from passing through.

– *Pharynx*: a passageway in the head and neck, part of the digestive and respiratory systems. The pharynx connects the nose to the mouth and throat.

– *Oropharynx*: anatomical region of the throat comprising: the tonsils (*responsible for trapping bacteria and viruses that may enter the body through the throat*), the soft palate, the back of the tongue and the back of the throat.

– *Larynx*: part of the respiratory system; it is the duct that connects the throat to the trachea. The vocal cords are two bands of muscle located in the middle of the larynx, which produce sounds and help us to speak. The larynx helps prevent food and liquids from entering the trachea.

3.6. Prostate disorders

In section 3.4.7, we discussed brachytherapy of the prostate using iodine-125 implants. In this section, we look at the general facts about prostate cancer and the principle of brachytherapy of the prostate using an iodine-125 implant.

3.6.1. *The prostate in the urinary tract*

The prostate is a gland of the male genital tract. It is an anatomical crossroad between the urinary tract and the male genital tract and is located below the bladder and in front of the rectum, surrounding the beginning of the urethra (Figure 3.18).

The rectum is the last part of the digestive tract before the anus. Its role is to store fecal matter before it is evacuated through the anus.

The bladder is an organ of the urinary system that stores urine until it is evacuated from the body. The function of the urethra depends on gender. In men, the urethra is the outlet for bladder urine and sperm; whereas, in women, only urine is evacuated via the urethra.

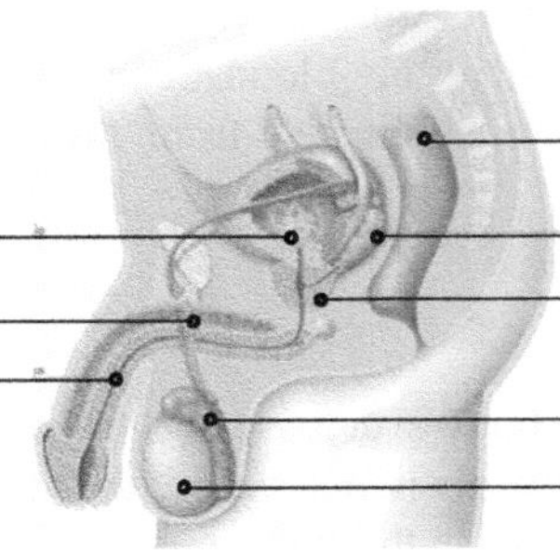

Figure 3.18. *The prostate in the urinary system. For a color version of this figure, see www.iste.co.uk/sakho/nuclearphysics3.zip*

The prostate plays a role in sperm production by producing a fluid called prostatic fluid. The seminal vesicles are involved in the production of seminal fluid, which is the main component of sperm and contains the elements needed to nourish and transport the spermatozoa during ejaculation. During ejaculation, the seminal fluid mixes with the spermatozoa, which come from the testicles and pass through the vas deferens.

The epididymis (*small funnel-shaped body on the upper edge of the testicle*) stores the spermatozoa so that they can reach full maturity. The mature spermatozoa travel through the vas deferens. Along the way, fluids produced by the seminal vesicles and prostate mix with the sperm to form semen. The testicles are male genital organs located in the scrotum at the base of the penis. They make sperm and produce testosterone, the hormone responsible for sexual desire.

There are three main conditions that can affect the prostate: adenoma (benign enlargement of the prostate), prostatitis (acute or chronic infection of the prostate) and cancer. These conditions are discussed in detail below.

3.6.2. *Benign prostatic adenoma or enlargement*

Definition

Prostate adenoma, benign prostatic hypertrophy (BPH) or benign prostatic hyperplasia is a benign pathology (in the sense of non-cancerous). It is a benign disease with no major risk to the individual, generally affecting men over the age of 50 (sometimes earlier) [BRO 22, HOS 23]. However, it can cause urinary problems due to the pressure exerted by urine on the urethra (see symptoms below).

Prostate adenoma is an enlargement of the prostate gland caused by an adenoma (benign tumor). It tends to develop from the glandular tissue surrounding the urethra; whereas, prostate cancer develops mainly in the posterior peripheral zone of the gland (Figure 3.19, section 3.6.3).

VOCABULARY CORNER.–

Adenoma: an increase in the size of the prostate gland usually linked to age, which can cause difficulty in urinating. This is also known as benign prostatic hyperplasia (BPH) [INS 10].

The causes, risk factors, symptoms, diagnosis and treatment of prostate adenoma are as follows.

Causes and risk factors

Furthermore, as far as is currently known, there are no causes of adenoma [HOS 23]. This disease results from a natural evolution of the central part of the prostate, a consequence of aging. This explains why it generally affects men over the age of 50. Furthermore, adenoma is not a risk factor for prostate cancer.

Symptoms and possible complications

Prostate adenoma presents two types of symptoms: irritative urinary disorders and obstructive urinary disorders. These symptoms may be accompanied by complications.

– Irritative urinary disorders occur mainly at night. They are characterized by a frequent, sudden and irrepressible need to urinate (that we cannot contain).

– Obstructive mictional disorders are due to a narrowing of the lumen (diameter) of the urethra. They result in difficulty in urinating, manifested by a reduction in the strength of the micturition stream, a sensation of incomplete emptying of the bladder and delayed drops at the end of micturition.

Acute urinary retention is the most striking and painful complication of prostatic adenoma. It has a sudden onset and sometimes marks the beginning of the disease. It is the blockage of micturition and therefore the bladder's inability to empty itself that causes the pain [BRO 22]. More rarely, the obstruction becomes more serious and leads to dilation of the bladder, which can no longer perform its reservoir and emptying functions correctly. This is known as chronic urinary retention (complete and painful inability to urinate). Patients can urinate, but without emptying completely, and may even experience overflow leaks. This situation is serious because it is a risk factor for bladder stones and renal failure [HOS 23].

VOCABULARY CORNER.–

Micturition: the act of urinating. Obstructive urination disorders include weakness of the stream, difficulty in initiating urination, a feeling of incomplete emptying; and, at the end of the course, acute urinary retention (AUR).

Irrepressible need: a need that cannot be contained or controlled.

Urinary incontinence: involuntary discharge of urine. Urinary incontinence can be either complete (loss of control of the sphincters that hold urine in the bladder) or incomplete, occurring during the day (when walking or exerting oneself) or at night, alternating with normal control.

Diagnosis

Diagnosis of prostate adenoma is based primarily on a clinical examination (including a digital rectal examination) and questioning of the patient. The *digital rectal examination* is a routine procedure used to assess the size, shape and consistency of the prostate. The interview is carried out by a doctor, who offers the patient a reference questionnaire, such as the IPSS questionnaire (see note). This consultation enables the severity of the urinary problems experienced to be assessed, as well as their impact on the patient's quality of life.

Additional tests may be carried out, such as the following:

– *Ultrasound* can be used to assess the size of the prostate and the existence of a median lobe (upper portion of the prostate), whose enlargement into the bladder obstructs the flow of urine.

– *Micturition flowmetry* is used to measure the flow of urine during micturition.

– *Cystoscopy* involves exploring the urethral canal and the inside of the bladder by inserting a small camera into the penis or urethra in women. It is used to clarify the diagnosis and guide the choice of surgical technique.

NOTE.–

IPSS stands for International Prostate Score Symptom. It is a set of questions and answers. The first question is: *Over the past month, how often have you had the feeling that your bladder was not completely empty after urinating?* Six answers are then given in six tick boxes arranged horizontally: (1) Never; (2) About 1 in 5 times; (3) About 1 in 3 times; (4) About 1 in 2 times; (5) About 2 in 3 times; (6) Almost always.

For more details, see: https://urofrance.org/fileadmin/medias/scores/score-IPSS.pdf.

Treatment

There are several possible treatments for benign prostatic hyperplasia, depending on the degree of discomfort caused by the prostate adenoma [HOS 23]:

– *Monitoring*: if urinary problems are minimal and do not cause any real discomfort, simple monitoring in the form of a check-up once or twice a year is sufficient.

– *Medical treatment*: when monitoring is not enough, or is no longer sufficient, and urinary problems have an impact on the patient's quality of life, drug treatment is used. If a single drug is ineffective, two drugs can be combined in a dual therapy.

– *Surgical treatment*: if the drug treatment loses its effectiveness or if there are repeated complications (acute urine retention, recurrent urinary tract infections, recurrent bladder stones, chronic urinary retention), surgery may be necessary. This involves removing prostate tissue (excision of the adenoma) so that urine can once again flow normally. This operation may be carried out under anesthetic (epidural or general) and requires hospitalization for 24–72 h. It should be noted that the main side effect of surgical treatment of benign prostatic hyperplasia is retrograde ejaculation (sperm flows back into the bladder instead of outwards).

3.6.3. *Prostatitis*

Definition

Prostatitis is an infection of the prostate gland. The term prostatitis covers two different conditions, acute prostatitis and chronic prostatitis.

An acute infection manifests itself suddenly and causes symptoms rapidly (as opposed to chronic diseases, which develop more slowly). *A chronic infection, on the contrary, is a long-term, progressive illness, often associated with disability and the threat of serious complications.*

There are generally four types of prostatitis [AND 22, ESC 22]:

– acute prostatitis of bacterial origin;

– chronic prostatitis of bacterial origin;

– chronic non-bacterial prostatitis or pelvic pain syndrome;

– asymptomatic prostatitis.

These different types of prostatitis are described as follows [SOC 23]:

– *Acute bacterial prostatitis* is caused by bacteria. It is said to be acute because its symptoms appear suddenly and can be severe. It is the least common type of prostatitis but the easiest to treat.

– *Chronic (long-term) bacterial prostatitis* is caused by bacteria and often persists for several months. Its symptoms are usually less intense than those caused by acute bacterial prostatitis and tend to disappear and then return. Chronic bacterial prostatitis may cause urinary symptoms such as burning or pain when urinating, but does not cause fever or chills. It appears to occur more often in older men with benign prostatic hyperplasia.

– *Chronic non-bacterial prostatitis/chronic pelvic pain syndrome (CP/CPPS)* is the most common type of prostatitis. It is a chronic condition not caused by bacteria. It causes pain or discomfort in the pelvis that may come and go.

– *Asymptomatic inflammatory prostatitis* causes no symptoms and requires no treatment. Doctors often diagnose it when they carry out examinations to check for the presence of other conditions, such as prostate cancer, or to determine the cause of infertility. To make this diagnosis, they need to detect white blood cells in the prostate or prostatic fluid.

Prostatitis can affect men of any age. However, it should be noted that bacterial prostatitis is more common in men under 35, while non-bacterial prostatitis is more common in men over 50. The *prevalence* of prostatitis is generally between 5% and 9%. According to one study, the annual incidence of prostatitis is around 5 out of every 1,000 men. 25% of people with genitourinary disorders also suffer from some type of inflammation of the prostate. Chronic prostatitis of non-bacterial origin accounts for *90–95% of cases of inflammation.* It is therefore the most common type. Acute or chronic prostatitis accounts for only 2–5% of cases. This means that bacterial inflammation of the prostate is not as common [ESC 22].

NOTE.– [BUI 21]

Bacteria are autonomous biological entities measuring between 1 and 10 microns. It is a unicellular organism without a nucleus, with a nuclear apparatus made up of DNA, a flagellum that enables it to move and a protective capsule. Bacteria can be aerobic, meaning that they need oxygen to develop, or anaerobic. They can reproduce by dividing and proliferating in different environments that are favorable to them. They can be found in the human body, as well as in water, soil or suspended in the air. Bacteria can be transmitted in a number of ways: through the air when a person coughs or sneezes, for example, or by ingesting water or food contaminated by a pathogen, through parenteral transmission, for example, with a syringe, or by direct contact with contaminated skin or surfaces.

Unlike bacteria, which are cells, viruses have a very simple structure. Smaller than bacteria, viruses measure between 20 and 500 nm. They consist of DNA or RNA (ribonucleic acid) surrounded by a protein structure called the capsid and sometimes, but not always, an envelope. Viruses are therefore classified according to the nature of the nucleic acid in the genome (RNA or DNA), the shape of the capsid and the presence or absence of an envelope, also known as the peplum.

Examples of bacteria and viruses

Escherichia coli is the most common bacterium found in the digestive tract of humans and warm-blooded organisms. Most strains are harmless. Some, however, can cause serious food poisoning. Among the seven most dangerous viruses are Ebola and SARS-CoV-2, the Covid-19 disease virus.

VOCABULARY CORNER.–

– Benign prostatic hyperplasia (BPH) is an increase in the size of the prostate caused by a proliferation (hyperplasia) of cells in the prostate.

The following are causes and risk factors. Symptoms, diagnosis and treatment of prostate adenoma are taken from Esculape [ESC 22].

Causes

Inflammation of the prostate can have both bacterial and non-bacterial origins. The causes vary according to the *type of prostatitis*.

– *Acute prostatitis of bacterial origin*: this can develop as a result of fecal bacteria, most often originating in the intestine. Examples include Escherichia coli, which is favored by poor hygiene, a deficiency in the immune system and factors such as stress, smoking, alcohol, dietary imbalances and a sedentary lifestyle.

– *Chronic prostatitis of bacterial origin*: this results from an infection of the prostate and is characterized by continuous healing and relapses. It is usually accompanied by recurrent urinary tract infections. Complications of an infection of the testicles (*epididymitis*) may also be a possible cause.

– *Chronic non-bacterial prostatitis*: specialists often fail to identify the *specific causes* of inflammation. As a result, symptoms appear, subside and then reappear.

Risk factors

The risk factors for prostatitis are: stress, anxiety, immune disorders, urinary tract malformations, constipation, hemorrhoids, professional or sporting trauma, multiple sexual partners, poor diet, excessive alcohol consumption, insertion of a catheter into the bladder, HIV infection, intense exertion with a full bladder, diabetes, severe constipation, immunosuppression and prolonged sexual arousal.

Sports such as cycling and horse-riding complete the list of risk factors for the development of prostatitis.

VOCABULARY CORNER.–

Immunodepression: a specific pathology reflecting a weakening of the immune system (the body's defense system).

Symptoms

The *symptoms of prostatitis* generally vary according to the type of inflammation, and can lead to complications if not treated in time. Prostatitis generally manifests itself as urinary irritation, pain in the lower abdomen, sexual problems and possible *reduced fertility*.

Diagnosis

To diagnose prostatitis, the GP first takes a *history*, asking the patient questions about their symptoms, general state of health, use of medication and *sexual habits*. Once the history has been taken, a full examination is carried out, including a digital rectal examination (DRE) to palpate the prostate and assess its consistency using a lubricated finger through the anterior wall of the rectum. In acute types of prostatitis, this *palpation of the prostate* causes intense pain. DRE can also be used to detect any enlargement of the prostate or the presence of adenomas. The doctor may also carry out a urine analysis using a test known as urine culture or *urine cytobacteriological examination*. This test is designed to determine the presence of signs of inflammation. It is also used to identify any pathogens responsible.

The doctor may also carry out the following tests:

– semen or seminal fluid analysis;

– blood tests (to detect any increase in prostate-specific antigen);

– urine flowmetry (for urinary reflux);

– bladder and prostate ultrasound;

– transrectal ultrasound;

– prostate biopsy.

In *recurrent and chronic forms*, it is essential to rule out obstructive factors that may predispose to prostatitis.

VOCABULARY CORNER.–

– *History* (return to the memory of the forgotten or repressed past). The *history* is an interrogation preceding the clinical examination. For the doctor, it consists of obtaining indexical information to help establish a diagnosis.

– *Biopsy*: a biopsy is a procedure in which a sample of tissue (or tumor) is removed for examination under a microscope. A biopsy is used to diagnose cancer or to determine whether or not an abnormality is cancerous.

Treatment of prostatitis [AND 22]

Treatment varies considerably, depending on the etiology.

– Acute bacterial prostatitis: in moderate (non-toxic) forms, patients can be treated at home with antibiotics, combined with bed rest, analgesics, laxatives and hydration.

– Chronic bacterial prostatitis: treated with oral antibiotics for at least six weeks.

– Chronic prostatitis/chronic pelvic pain syndrome: treatment is difficult and often unsatisfactory.

– Asymptomatic inflammatory prostatitis: this condition requires no treatment.

VOCABULARY CORNER.–

– *Analgesic*: medicine that reduces or eliminates pain without treating the cause.

– *Laxative*: product that speeds up intestinal transit and softens stools. It is used to combat constipation.

– *Etiology*: all the causes of a disease.

3.6.4. *Prostate cancer*

Benign prostatic hyperplasia tends to develop from the glandular tissue surrounding the urethra; whereas, prostate cancer develops mainly in the posterior peripheral zone of the gland (Figure 3.19).

Adenomas tend to develop from the glandular tissue surrounding the urethra; whereas, prostate cancer develops mainly in the posterior peripheral zone of the gland (Figure 3.19) [YOU 20, HOS 23].

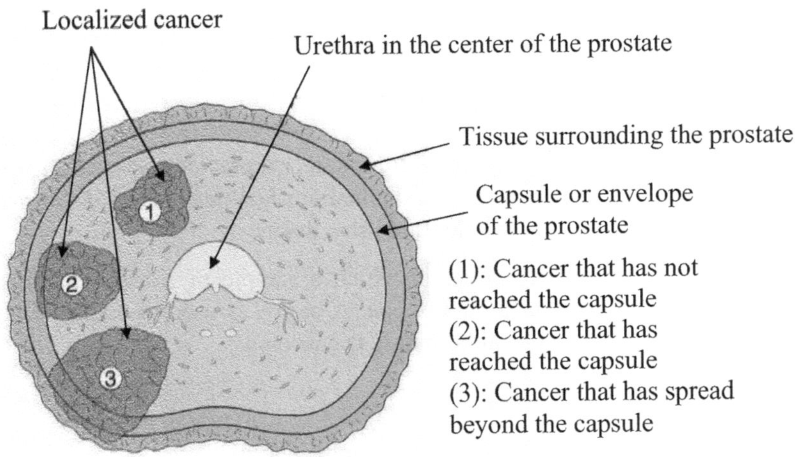

Figure 3.19. *Cross-section of the prostate showing the different stages of development of prostate cancer and the tumor. Source: https://www.cancer-de-la-prostate.fr/comprendre-cancer-prostate/. For a color version of this figure, see www.iste.co.uk/sakho/nuclearphysics3.zip*

Prostate cancer develops from the prostate cells that line the small glands within the prostate. The disease usually goes completely unnoticed for several years, then symptoms gradually appear as the tumor grows.

Prostate cancer is usually adenocarcinoma. Prostate adenocarcinoma is the most common non-dermatological cancer in men over the age of 50 in the United States. In the United States, for example, there were around 248,530 new cases and around 34,130 deaths in 2021. The incidence increases with each decade of life. Autopsy studies show the presence of prostate cancer in 15% to 60% of men aged between 60 and 90, with the incidence increasing with age. The lifetime risk of developing prostate cancer is 1 in 6. The median age at diagnosis is 72, and over 75% of prostate cancers are diagnosed in men over 65. The risk is higher in black men [CHA 22]. Every year, around 12.000 men die from prostate cancer in the United Kingdom, and there are 50,000 new cases per year in France, with 8,000 deaths [HUR 22]. In Senegal, there are 880 new cases of prostate cancer and 504 deaths caused by this pathology, considered to be the fourth most common cancer [DIE 22]. In percentage terms, estimates for 2019 show that the most common cancers in Senegal are cervical cancer (17.8%), breast cancer (16.7%), liver cancer (10.2%), prostate cancer (9.1%) and stomach cancer (5.2%) [BRI 19]. Recent studies have shown that Senegal records 11,317 new cases of cancer and around 8,000 deaths each year [NDA 23].

VOCABULARY CORNER.–

Screening: in medicine, screening consists of looking for one or more "risk" diseases or abnormalities in individuals from a given population. These investigations may or may not be followed by medical consultations, clinical examinations and treatment.

3.6.5. *Symptoms and diagnosis of prostate cancer*

As the prostate is an organ of the urinary and reproductive systems, the first symptoms of a prostate tumor are closely linked to urinary and sexual functions. These symptoms can often be linked to a benign prostate tumor (prostate adenoma) rather than prostate cancer. Prostate cancer can also cause general symptoms (general weakness, weight loss, etc.), but these are usually late signs, occurring in the later stages of the disease.

The first warning sign is an abnormally frequent and/or urgent need to urinate. Urination may also be difficult or even painful, accompanied by burning or discomfort. Abnormal sensations often accompany urination, in particular, the feeling that the bladder has not been completely emptied after urinating. Urinary incontinence is also a warning sign that should not be ignored.

Hematuria, the presence of blood in the urine, is also an early symptom of prostate cancer, as well as bladder or kidney tumors (which are far rarer). The presence of blood in the semen is also suggestive of prostate cancer. Finally, pain in the pelvic region and/or lower back, as well as a feeling of discomfort or heaviness, are sometimes observed [SCH 22].

The symptoms of advanced prostate cancer are [YOU 20, VID 23]:

– difficulty urinating: frequent need to urinate, low flow rate, difficulty producing a constant stream;

– blood in the urine or semen;

– painful ejaculation, sometimes with erectile dysfunction;

– pain in the pelvis, hips or upper thighs;

– irritation of the rectum;

– pain in the lower abdomen or lower back;

– erectile dysfunction.

When prostate cancer has invaded other organs, the patient may suffer from fatigue, nausea, weight loss, loss of appetite, bone pain, etc.

Given the importance of the cancer problem in Senegal, our work in radiotherapy focuses in particular on the study of the dose to distant organs (bladder, rectum, etc.) from the tumor (prostate cancer). The aim is to assess the risk of radiation-induced secondary cancers in order to prevent them and improve dose calculation systems (TPS) [NGU 23].

Diagnosis

However, most cases remain asymptomatic. Cancer can be detected by screening: either by checking the PSA "prostate-specific antigen" level through a blood test or by digital rectal examination. The diagnosis is confirmed by a series of biopsies. A biopsy involves removing a fragment of tissue or organ (often called a core) using a special needle, so that it can be analyzed under a microscope for histopathology. This analysis reveals the geometry of the cancer cells (shape and size), enabling us to assess their nature, extent and degree of aggressiveness. This procedure is most often guided by ultrasound imaging. Clinicians have access to 2D ultrasound visual feedback [YOU 20].

3.6.6. *Treatment methods for prostate cancer*

The main treatment modalities for prostate cancer are [INS 10]: surgery, radiotherapy (external radiotherapy and brachytherapy), active surveillance (which allows treatment to be deferred) and hormone therapy. Chemotherapy may also be used on rare occasions. Other techniques are currently being evaluated: treatment with high-intensity focused ultrasound and cryotherapy.

Treatments may be used alone or in combination. Depending on the case, their objectives may be [INS 10]:

– to eliminate the tumor or metastases;

– to reduce the risk of recidivism;

– to slow the development of the tumor or metastases;

– to treat the symptoms of the disease to ensure the best possible quality of life.

VOCABULARY CORNER.–

Cryotherapy: local treatment that destroys cancer cells using intense cold. Guided by an ultrasound scan, the doctor inserts needles into the prostate through the skin between the testicles and the anus, freezing the tumor at between -40°C and -60°C. The cold causes the cancer cells to rupture and die. This is also known as cryosurgery.

In the following section, we study the special case of brachytherapy (internal radiotherapy) of the prostate using iodine-125 implants. Finally, we will briefly study the special case of prostate chemotherapy used in various oncology treatment centers such as the CICD (Centre International de Cancérologie de Dakar; see Figure 1.10, Chapter 1) is on duty.

3.7. Principle of prostate brachytherapy using an iodine-125 implant

3.7.1. *Decay diagram for radioiodine-125*

Of the radioisotopes mentioned above, iodine-125 sources are most commonly used in brachytherapy because of their usefulness in treating many types of cancer, such as vaginal, cervical, eye and prostate cancer. Due to the specific nature of iodine-125, it is important to explain in detail the origin of the electromagnetic radiation produced by this radioelement.

Iodine-125 decays by electron capture to the ground level of tellurium-125 via the excited level of 0.0355 MeV. The de-excitation of the ^{125}Te daughter nucleus towards the fundamental level is then accompanied by the emission of a 0.0355 MeV photon (Figure 3.20).

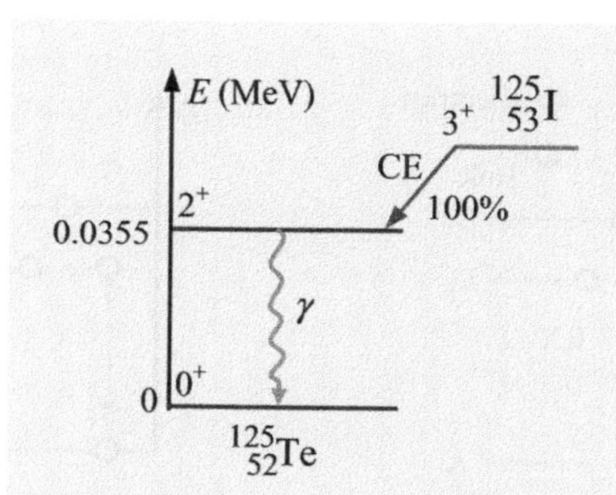

Figure 3.20. *Decay energy diagram of iodine-125 by electron capture (EC). For a color version of this figure, see www.iste.co.uk/sakho/nuclearphysics3.zip*

Note that the rearrangement of the electron pattern can be accompanied by the emission of *Auger electrons* by a process known as the *Auger effect*. It follows that electron capture is accompanied by a competition of X-ray photon and Auger electron emissions.

When the nucleus of ^{125}I captures an electron from the electron chain, an electron is absorbed by a proton in the nucleus. A neutron and a neutrino are formed. As the neutrino is not a component of the nucleus, it is ejected without emitting any detectable radiation. The gap created by the capture of an electron from an internal orbital very close to the nucleus is compensated for by the rearrangement of the electron pattern. This rearrangement is then accompanied by the emission of X-ray fluorescence and Auger electrons. This radiation is characterized by energetic line spectra. The short half-life of ^{125}I (59 days) means that the radiation emitted after one year can be considered negligible (1/100). ^{125}I emits radiation with an average energy of 35 keV. This radiation has very low penetration and is absorbed in the first few millimeters of tissue. The various X-rays and gamma rays are indirectly responsible for dose deposition in matter. Due to their very low energy, Auger electrons do not contribute to the dose; they are stopped before they enter the medium by the encapsulation of the sources [THI 07].

Let us briefly describe the mechanism of competition between the emission of X photons and Auger electrons [SAK 22]. Suppose an electron from the K layer is captured by a nucleus. The electron gap thus created can be occupied by one of the electrons from the upper layers L, M, N, etc. This results in an electronic rearrangement with the emission of X photons characteristic of the emitting atom by a process called *electron–hole recombination*. This process competes with the emission of an Auger electron (e^A) by a process called the *Auger effect*. Figure 3.21 shows atomic de-excitation by X-ray or Auger electron emission, following K-layer electron capture [SAK 22].

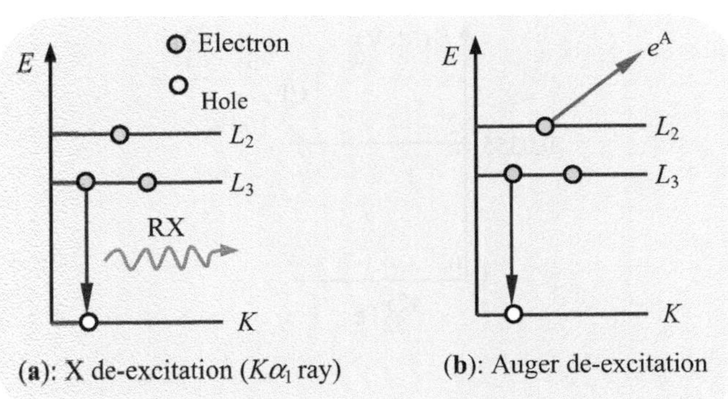

Figure 3.21. *Atomic de-excitation process by emission of X-rays or of an Auger electron by rearrangement of the electron chain following capture of a K-layer electron. For a color version of this figure, see www.iste.co.uk/sakho/nuclearphysics3.zip*

NOTE.–

Internal conversion is a nuclear de-excitation process that generally competes with γ de-excitation. It is a process in which the excitation energy of a nucleus is transferred directly to an electron in the atomic nucleus. The electron with binding energy W is then expelled from the atom. This produces the emission of an electron called an *internal conversion electron* (e^C) with a specific kinetic energy. This competition is particularly important at low excitation energies [SAK 22]. The internal conversion electron creates an electron gap in one of the inner layers of the atom, resulting in the emission of X-ray photons characteristic of the emitting atom, in competition with the emission of Auger electrons.

3.7.2. Definition of brachytherapy: LDR, HDR and PDR brachytherapy modes

Brachytherapy (from the Greek word "brachy", meaning "short distance") is a local cancer treatment whose aim is to destroy cancer cells by means of rays produced by a radioactive substance placed inside an organ or in contact with it. This implant, in the form of seeds or wires, may be temporary or permanent. Only certain organs can be treated by brachytherapy: uterus, prostate, breast, throat, mouth [ICN 10].

There are two types of brachytherapy, depending on whether or not the radioactive sources used are removed: brachytherapy using temporary permanent implants (the sources are left in place for a few hours or days, and then removed) and brachytherapy using permanent implants that are left in place permanently.

There are two forms of prostate brachytherapy [YOU 20]:

– High-dose-rate brachytherapy: in this case, the radioactive sources are left for several minutes to several hours, depending on the dose rate used, and then removed.

– Low-dose-rate brachytherapy: in this case, the radioactive sources injected via needles remain permanently implanted in the prostate.

As an internal radiotherapy technique, brachytherapy therefore involves the precise positioning of radiation sources directly on the area of the cancerous tumor, unlike external radiotherapy, where high γ or X-rays from outside the body are directed at the tumor to be treated. The radioelements used in brachytherapy include iodine-125 (^{125}I. 59.4 d) and iridium-192 (Ir. 74 d) [ZIM 06, THI 07, COM 18]. This can be a surgical implant (porcelain iodine-125 bead in contact with the prostate) or a wire (iridium-192) guided in a catheter (probe), whose tip leads to the tumor site to be treated [COM 18].

One of the important features of brachytherapy is that a dose (expressed in gray per hour $Gy \cdot h^{-1}$) can be delivered in less time than with other radiotherapy techniques. This can then help to reduce the possibility of cancer cells surviving, dividing or growing between each radiotherapy session. For localized irradiation, brachytherapy involves bringing an adequate quantity of radioactive material as close as possible to the cancerous tissue. For this irradiation to be sufficient and effective, without causing undesirable effects on neighboring healthy tissue, several conditions must be met. The radiation emitted must have a short range in order to control energy deposition in the tumor and neighboring tissues. In addition, the radionuclide must remain in contact with the tissue to be destroyed for a sufficiently long time. Its concentration must also provide enough energy to destroy all the tumor cells [THI 07].

There are also different types of brachytherapy, classified according to the dose rate used. These are low-dose-rate (LDR) brachytherapy, high-dose-rate (HDR) brachytherapy and pulsed-dose-rate (PDR) brachytherapy.

– LDR is characterized by a dose rate between $2\ Gy \cdot h^{-1}$ and $12\ Gy \cdot h^{-1}$.

– HDR corresponds to a dose rate greater than $12\ Gy \cdot h^{-1}$.

– PDR involves delivering short pulses of radiation, usually once an hour, to simulate the overall throughput and efficiency of LDR processing.

In practical terms, LDR brachytherapy is commonly used, for example, for cancers of the oral cavity and sarcomas. HDR brachytherapy is used to treat cervical, esophageal, lung, breast and prostate cancers. PDR brachytherapy is used to treat, for example, cancers of the head and neck.

Brachytherapy can be used on its own or in combination with other treatments such as surgery, external radiotherapy and chemotherapy. The cure rate for cancer using brachytherapy is comparable to that of surgery or external radiotherapy. In addition, brachytherapy has the advantage of being associated with a low risk of undesirable and serious side effects [FER 07, FRA 07].

NOTE.–

i) The mouth, or buccal cavity, is the first segment of the digestive tract, and is also used for breathing and phonation. It is lined on the inside by a protective mucous membrane and opens through the lips. Sarcomas are malignant tumors that develop from the body's supporting tissues or soft tissues, such as muscle, fatty tissue, cartilage, blood or lymphatic vessels, nerves or bone. The type of sarcoma depends on the type of cells that give rise to it.

ii) Sarcomas are malignant tumors that develop from the body's supporting tissues or soft tissues, such as muscle, adipose tissue, cartilage, blood or lymphatic vessels, nerves or bone. The type of sarcoma depends on the type of cells that give rise to it [DÉP 22].

3.7.3. *Principle of implantation of prostate brachytherapy equipment*

In brachytherapy using permanent implants, the seeds of iodine-125 are inserted using hollow needles, either manually or automatically, under ultrasound control. The needles are removed as the radioactive material is implanted. At the end of the implantation, the patient is taken to the recovery room. An imaging test, such as an X-ray, may be carried out to check the quality of the iodine seeds implanted in the prostate. This may be supplemented by other imaging tests.

In temporary implant brachytherapy, the needles placed in the prostate are connected to a source projector, which usually contains an iridium micro-source. This micro-source automatically moves through each needle according to the dosimetry program. Depending on your case and the habits of your medical team, one to three sessions can generally be planned. When several sessions are planned, they may be spread over several days [INS 22]. Figure 3.22 shows the principle of implanting brachytherapy equipment in the treatment of prostate cancer.

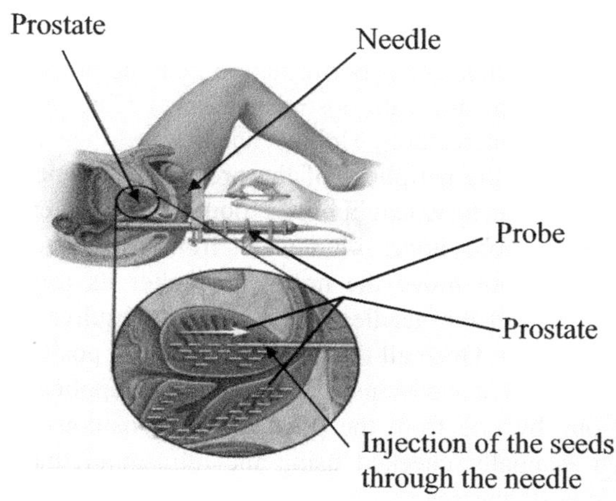

Figure 3.22. *Principle of implantation of brachytherapy equipment for prostate cancer [YOU 20]. For a color version of this figure, see www.iste.co.uk/sakho/nuclearphysics3.zip*

Prostate brachytherapy, involving the permanent implantation of radioactive grains of iodine-125, offers patients with localized prostate cancer an attractive alternative to surgical treatment [HER 14]. In the case of prostate cancer, which is most common in men, permanent HDR brachytherapy or temporary HDR brachytherapy is a novel approach to treating prostate cancer. However, this approach, also known as boost therapy, is still used less frequently than permanent seed implantation. Temporary HDR brachytherapy is used when it is necessary to administer an additional dose to the patient (while sparing the surrounding tissues from radiation exposure), compared with external radiotherapy, depending on the type of prostate tumor [GAL 04].

Iodine implant brachytherapy is performed in three consecutive stages [YOU 20]:

First planning stage

In general, a pre-operative examination guided by ultrasound imaging, usually accompanied by an MRI scan, is carried out about 10 days before the brachytherapy procedure. This examination is used to prepare the planning by determining the volume of the prostate and calculating the number of radioactive seeds to be placed to produce the desired dose. This is the first stage of planning.

Second stage

This stage is carried out on the day of the operation. On this day, an ultrasound probe is used to guide the procedure and a template with a matrix of holes every 5 mm is used to insert the needles. First of all, the patient is put under general anesthetic and an ultrasound scan is performed to plan the procedure, specifying the positions of the needles in line with dose constraints on the prostate and organs at risk (prostate, urethra and rectum). Hollow needles are then inserted one by one through the grid around the periphery of the prostate. Each needle then follows a theoretical horizontal trajectory, leaving little room for maneuver when it encounters an obstacle, such as the pubic bone, if it deviates from its trajectory. In the event of a blockage, the surgeon can move the needle and alter its trajectory by applying movements to the base of the needle. This technique requires a certain amount of skill from the practitioner. Once all the needles have been positioned correctly, they are loaded with radioactive seeds and injected using an applicator. The needles are then withdrawn one by one from the prostate. The positions of the seeds in the prostate are then manually checked using the ultrasound image. Following this check, the positions of the seeds are updated in the dose calculation software. A second round of seed implantation is carried out, based on the updated dose. This time, the needles are inserted into the central positions of the prostate. Once all the needles have been correctly positioned, the seeds are injected in the same way. At the end of the second stage, an X-ray is taken in the OR to visualize all the seeds that have been implanted. The prostate, however, is not visible on these images.

Third stage

This stage enables the location of the final grains implanted to be determined once the edema associated with the procedure has subsided. This is done by CT scan one month later. This monitoring stage documents the progress of the brachytherapy. No further corrective action is possible at this stage. Each needle delivers between 2 and 6 seeds. These seeds may be free or connected. This is known as a seed train. This is made up of seeds connected by connectors and spaced by non-metallic spacers. The free seeds are injected one by one. This makes the injection procedure more cumbersome, and can lead to inaccuracies in implantation due to seed migration. The use of a train of seeds minimizes implantation inaccuracies by injecting only once per needle. However, it leaves less freedom in terms of seed position.

3.7.4. *Prostatic chemotherapy*

Definition

Chemotherapy (from the Greek chemo = chemical and therapy) is a method of systemic cancer treatment.

Generally speaking, a distinction is made between systemic treatments, which are frequently used in the fight against cancer, and local treatments (treatment of very localized areas) such as surgery and radiotherapy. In this way, chemotherapy acts throughout the human body.

Chemotherapy is a treatment for cancer using drugs that kill or weaken cancer cells. It is a general treatment aimed at destroying cancer cells or preventing them from multiplying throughout the body [INS 10].

Depending on the type of tumor and the type of cancer (breast cancer, prostate cancer, cancers of the ENT sphere, uterine cancer, esophageal cancer, etc.), chemotherapy may be neoadjuvant (as the first treatment, before surgery or radiotherapy) or adjuvant (after surgery). It may also be concomitant with radiotherapy and be carried out at the same time as the latter: this is known as radiochemotherapy or chemoradiotherapy, with the aim of potentiating the effect of radiotherapy [DAR 21].

Mode of operation

Chemotherapy treatment acts on cancer cells, regardless of their location in the human body or their visibility on imaging. Its action slows and blocks the growth and spread of cancer cells in the blood. This reduces the size of the tumor, eliminates cancer cells that remain in the body after surgery or radiotherapy and reduces the risk of recurrence of the cancer. Chemotherapy can act on a primary tumor or on metastases further away from the initial tumor site. However, the drugs used for chemotherapy also have consequences for the whole body, depending on the patient and the combination of drugs. This is reflected in the possible appearance of side effects [CEN 22].

How the protocol works

Chemotherapy involves the *intravenous* (injection) or *oral* (tablet) administration of one or more chemotherapeutic drugs to eliminate tumor cells in the body. Whatever the method of administration, the drugs used are carried by the blood throughout the body. However, the injectable route remains the most common method of administration.

Several cycles of chemotherapy are planned as part of the therapeutic management of a given cancer. Repeated injections tend to irritate or even damage patients' veins. In addition, infusions are carried out via a flexible tube (catheter) and are associated with a number of constraints. IV injections are carried out in a chemotherapy facility, usually on an outpatient basis (which does not require hospitalization, and therefore involves scheduled travel to the treatment center). In addition, the catheter needs to be cared for to avoid the risk of infection. For this reason, oral chemotherapy to treat cancer is booming, and chemotherapeutic drugs are increasingly available in tablet or capsule form, that can be swallowed.

There is a rest period between each cycle of chemotherapy. This allows the body to rest and reduces any unpleasant side effects. A cycle generally lasts 3–4 weeks. The number of cycles depends on the drugs to be administered, the type of tumor and the side effects observed. Before each chemotherapy session, a complete examination is carried out, including a blood test and an interview. This enables the patient's state of health to be checked and ensures that there are no contraindications to the new course of chemotherapy [DAR 21].

Frequently used drugs

The two drugs used in chemotherapy are *docetaxel* and, if docetaxel fails, *cabazitaxel* (Figure 3.23). These two *antitumor agents* are administered in *10 courses* spaced *21 days apart*. They act by *slowing* or *blocking cell division*.

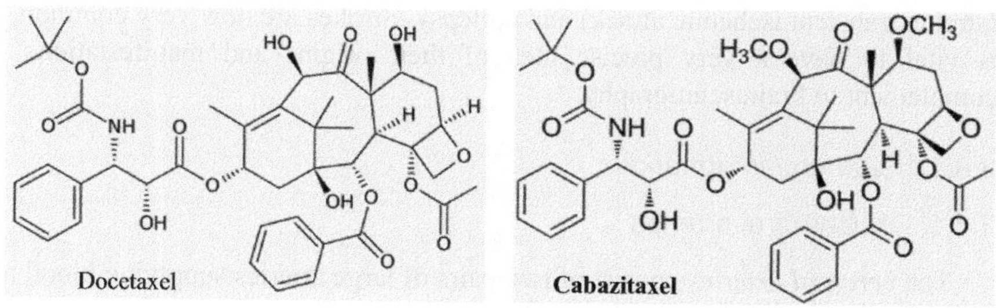

Figure 3.23. *Structures of docetaxel ($C_{43}H_{53}NO_{14}$) and cabazitaxel ($C_{45}H_{57}NO_{14}$) molecules*

As *cancer cells divide* particularly rapidly, they are the *first to be affected* by anti-tumor agents. However, *healthy* dividing cells are also *destroyed by antitumor agents*, giving rise to the *side effects of chemotherapy* [DAR 21].

Side effects of chemotherapy

The side effects of chemotherapy *depend on the type of cancer and the patient being treated. Their intensity and nature depend on the individual patient, and are not related to the efficacy of the treatment.* The most common side effects are digestive problems (diarrhea, vomiting, nausea, constipation, etc.) and damage to the mucous membranes of the mouth, anus and digestive tract; hair loss and skin fragility (redness, dryness, discomfort, etc.) may also occur, as well as changes in blood cells that can cause fatigue, shortness of breath, increased susceptibility to infections and coagulation problems. *Although these side effects can be frightening, they are now very well managed by modern medicine and should not keep patients away from this effective cancer treatment* [DAR 21].

In the case of prostate cancer, the common side effects of the chemotherapeutic agents administered are nausea and vomiting, diarrhea, low blood cell count (bone marrow aplasia), fatigue, hair loss, sore mouth and throat, and loss of appetite [SOC 21].

3.8. Appendices

3.8.1. *Stroke*

In section 2.5.3, we saw that cerebral scintigraphy with HMPAO or ECD examines the blood flow in the brain highlighting the areas of the brain that are more irrigated. This can be useful for a proper diagnosis of cognitive disorders, stroke

(stroke, transient ischemic attack) and epilepsy. Strokes are now very common, so it is vital to have a very precise idea of their origins and manifestations as a complement to brain scintigraphy.

3.8.1.1. *General information*

3.8.1.1.1. Cerebral arteries

The *cerebral arteries* consist of two pairs of large arteries supplying blood to the brain:

– The internal *carotid arteries*: these carry blood from the heart through the front of the neck.

– The *vertebral arteries*: these carry blood from the heart through the back of the neck.

There are three carotid arteries on each side of the neck:

– The larger-caliber common carotid artery (CCA): this artery, located in the lateral region of the neck, drains blood to the face, skull and brain. The CCA then divides into two arteries: the external carotid artery (ECA), which drains blood to the face and scalp, and the internal carotid artery (ICA), which drains oxygenated blood to the brain, eyes and inner ears.

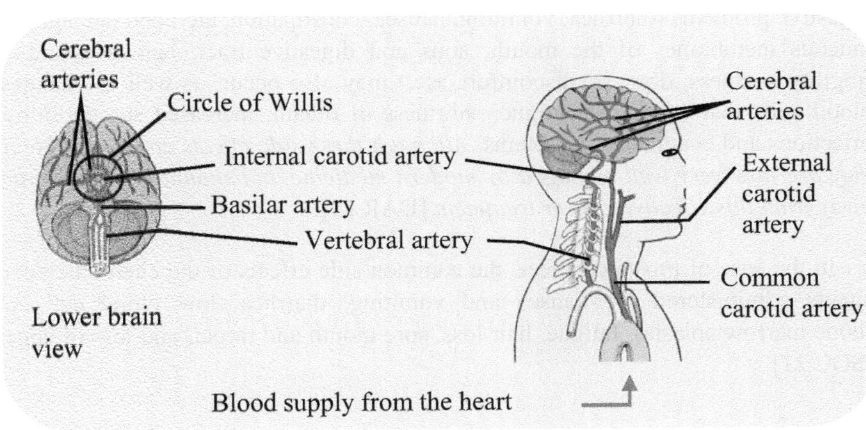

Figure 3.24. *Cerebral arteries and arterial circle of the brain or Willis polygon. For a color version of this figure, see www.iste.co.uk/sakho/nuclearphysics3.zip*

In the skull, the vertebral arteries merge to form the basilar trunk (at the back of the head). The internal carotid arteries and the basilar trunk divide into several branches, including the cerebral arteries. Some branches join to form a circle of arteries known as the *arterial circle of the brain,* the *circle of Willis* or the *Willis polygon* (in honor of the English physician and neurologist Thomas Willis (1621–1675), a precursor of neuropathology), which connects the vertebral arteries and the internal carotid arteries. Other arteries branch off from Willis' polygon like a roundabout (Figure 3.24). The basilar artery, which passes through the gutter formed by the central relief of the basilar trunk bridge, helps to supply blood to the brain.

3.8.1.1.2. *Obstruction of cerebral arteries: stroke*

Obstruction of the large arteries that supply blood to the brain may cause no symptoms or only a minor *stroke*. However, for the same obstruction, some people will have a serious *ischemic stroke*. *Collateral arteries* are part of the explanation.

Collateral arteries include the Willis polygon and the connections between arteries originating from the polygon. Some people are born with large collateral arteries, which can protect them from stroke. Therefore, when one artery is blocked, blood flow continues through a collateral artery, sometimes preventing a stroke. Other people are born with small collateral arteries. Collateral arteries that are too small may not be able to provide sufficient blood flow, leading to a stroke.

The body can also protect itself against stroke by building new arteries. When blockages develop slowly and progressively, new arteries can be created over time to continue supplying blood to the affected area of the brain and thus prevent a stroke. If a stroke has already occurred, the development of new arteries can prevent a second stroke (but cannot cure the lesions already present) [CHO 20b].

3.8.1.1.3. *The different types of stroke*

A stroke is a sudden vascular neurological deficit caused by an *infarct* or *hemorrhage* in the brain. In most cases, however, the causes are internal, linked to age, diet or lifestyle. Stroke symptoms vary greatly from case to case, depending on the nature of the stroke, the area of the brain affected and the size of the brain lesion. A stroke is also known as a *cerebral attack* or *cerebral infarction*. In the event of a stroke, symptoms persist and a brain lesion is most often visible on X-ray examinations (CT scan, MRI). There are two types of stroke: ischemic and hemorrhagic:

– *Ischemic stroke*: this is typically caused by blockage of a cerebral artery, especially a branch of one of the internal carotid arteries (Figure 3.1). The *clot* may form locally in a small artery or come from the heart, or from a lesion in the wall of one of the large *cervical arteries* (carotid and vertebral arteries). The result is a reduction in blood supply to an area of the brain. This is known as an ischemic stroke (Figure 3.25). More than 80% of strokes are ischemic [BÊM 19, CHA 19, CHO 20b, HÔP 21, CON 21].

Symptoms of ischemic stroke appear suddenly. They may include muscle weakness, paralysis, abnormal sensation or lack of sensation on one side of the body, slurred speech, confusion, blurred vision, dizziness and loss of balance and coordination [HÔP 21]. As a result, brain cells are deprived of blood. Most brain cells die if deprived of blood for 4.5 h [CHO 20b].

The diagnosis and treatment of ischemic stroke are as follows [CHO 20b, HÔP 23b]:

Diagnosis [CHO 20b]

The doctor can usually identify which cerebral artery is blocked on the basis of symptoms. For example, weakness or paralysis in the left leg suggests a blockage in the artery that supplies the area on the right side of the brain, which controls movement of the muscles in the left leg.

A CT scan is usually performed first. A CT scan can also differentiate between ischemic stroke and hemorrhagic stroke, brain tumors, abscesses and other structural changes. The doctor will also measure *blood sugar* levels, as low blood sugar (hypoglycemia) can cause similar symptoms. If available, a *diffusion-weighted MRI scan* (see vocabulary corner below), which can often detect ischemic stroke within minutes of the onset of symptoms, may then be performed.

Treatment [HÔP 23b]

Vital functions (oxygen, blood pressure, body temperature) and *neurological functions* (speech, strength, sensitivity) are monitored immediately. Care adapted to the victim is then initiated, depending on the origin of the stroke and its causes.

Thrombolysis consists of injecting a powerful drug into a vein or artery to dissolve the clot obstructing one of the vessels. This treatment is only possible within four and a half hours of the onset of symptoms, depending on the victim's condition.

Thrombectomy, or more precisely *mechanical thrombectomy*, involves removing the blood clot by inserting a special probe into the blocked artery to catch, break up and remove the clot in order to restore blood flow. The clot is then extracted using a stent (a small cylinder often used to dilate coronary arteries), which grips it between its meshes. The stent is finally removed.

Drugs, such as anti-platelet agents that prevent the formation of clots (for example, aspirin) or anticoagulants, are also administered depending on the case.

– *Hemorrhagic stroke*: this is the rupture of a blood vessel, leading to hemorrhage (Figure 3.25). Blood spills out and damages the surrounding brain tissue. The location and size of the hematoma determine its severity. This is known as hemorrhagic stroke, cerebral hemorrhage or *cerebral hematoma*. It is estimated that 20% of strokes are hemorrhagic [BÊM 19, CON 21]. Figure 3.25 shows the different origins of ischemic and hemorrhagic stroke. The suffering of the affected cells is insufficient to constitute an infarct (cell death). As a result, cerebral imaging is normal and there are no after-effects. This type of stroke does, however, give rise to fears of a subsequent ischemic event characterized by an infarct with permanent lesions. This is a diagnostic and therapeutic emergency.

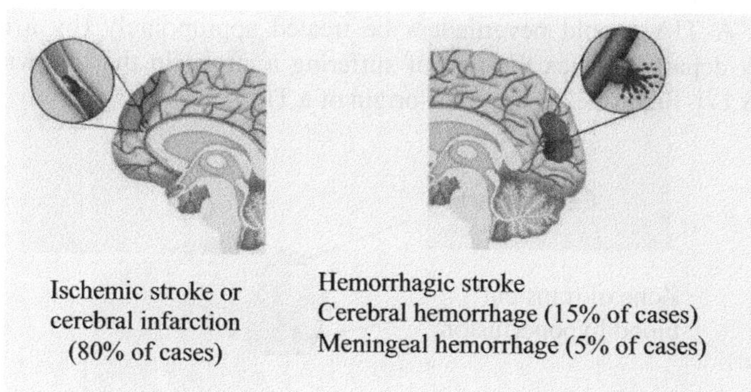

Figure 3.25. *Origins of the different forms of stroke (see: http://recap-ide.blogspot.com/2014/11/accident-vasculaire-cerebral-avc.html). For a color version of this figure, see www.iste.co.uk/sakho/nuclearphysics3.zip*

As shown in Figure 3.25, hemorrhagic stroke is caused by cerebral hemorrhage in 15% of cases and meningeal or subarachnoid hemorrhage (SAH) in 5% of cases. The characteristics of cerebral hemorrhage and meningeal hemorrhage are as follows [HÔP 23a]:

A meningeal hemorrhage is an *intracranial bleed* when blood erupts into the *meninges*, which surround the brain and spinal cord. The blood then mixes with the cerebrospinal fluid contained between the meningeal envelopes. A meningeal hemorrhage is revealed by an extremely violent and sudden headache, and may be complicated by loss of consciousness or even the onset of a coma. The hemorrhage usually affects a young, previously healthy patient. Immediate hospital care is essential.

Cerebral hemorrhage is bleeding that occurs suddenly when a small vessel ruptures, creating a *hematoma* inside the brain. In most cases, there is a *violent headache*, sometimes accompanied by vomiting. Other symptoms depend on the location of the hematoma in the brain. The main ones can be: paralysis on one side of the body (hemiplegia), sensitivity problems on one side of the body, coordination problems and clumsiness on one side of the body, language problems (aphasia), dizziness and vision problems.

The ischemic event may also be transient. This is known as a *transient ischemic attack* (TIA).

Like ischemic stroke, TIA is caused by the presence of a blood clot that temporarily or transiently obstructs cerebral blood flow [ROS 17, CHA 19, HÔP 21]. A TIA should nevertheless be treated appropriately (by a call to the emergency department), as the risk of suffering a stroke in the following days is high [ROS 17]. Figure 3.26 shows the origin of a TIA.

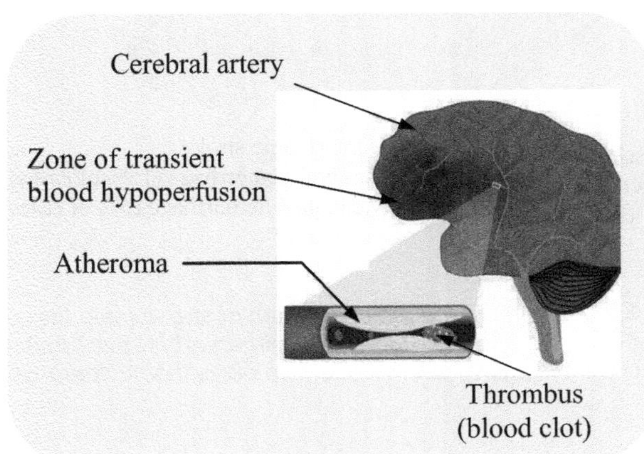

Figure 3.26. *Mechanism of transient ischemic attack (TIA) [ROS 17]. For a color version of this figure, see www.iste.co.uk/sakho/nuclearphysics3.zip*

TIA encompasses three concepts defined as follows [CHA 19]:

– it is an *attack*, i.e. an unforeseeable event;

– it is *ischemic*, which means that it is caused by poor oxygenation of the brain, most often due to an interruption in blood circulation;

– it is *transient*, which means it usually recovers.

A TIA causes a temporary disturbance of *brain function* (neurological disorders) that lasts as long as the brain's oxygen supply is inadequate. Neurological disorders depend on the area of the brain affected. The deeper it is, the more extensive it is, the greater and longer-lasting the disturbance. This is why a TIA is characterized by:

– *localized ischemia* (suffering of the territory without lesion);

– *complete* or *incomplete interruption* of circulation in the corresponding vessel, lasting less than 24 h;

– an *interruption* that only causes transitory signs, leaving no after-effects.

However, a TIA can foreshadow a stroke, with permanent deficits and serious after-effects. This is why, all TIAs require immediate medical attention. TIAs disappear as quickly as they appear:

– it leaves no after-effects;

– symptoms are highly varied, ranging from speech, hearing, sight and motor disorders. These symptoms:

- usually appear after the age of 60,

- do not last more than a few minutes,

- disappear completely within 24 h.

After 24 h, a TIA is considered a stroke, the consequences of which can be severe.

In addition, emergency signs are generally quite sudden and worrying. These signs are as follows [CHA 19]:

– difficulty speaking (*dysphasia*) or complete inability to speak (aphasia);

– asymmetry of the mouth or half of the face: this can also be facial paralysis from a different cause, but it also needs to be treated quickly;

– loss of sight in one eye;

– numbness or complete paralysis of half of the body;

– sudden weakness of the hand;

– a fall on one side due to sudden weakness of one leg.

The modalities for the diagnosis and treatment of TIA are as follows [ROS 17]:

Diagnosis

Biological tests, *electrocardiogram* (ECG), *cervical echo-Doppler*, brain scan and sometimes *carotid arteriography* are used to make the diagnosis. Today, however, it is MRI, above all, that enables a precise diagnosis to be made in the first few hours, or even minutes. It should be carried out as a matter of urgency in the presence of signs suggestive of cerebral ischemia.

Treatment

A patient suffering a TIA needs to be admitted to hospital. A CT scan is performed, as well as an MRI and a biological work-up, an electrocardiogram, a cervical and, if possible, transcranial echo-Doppler. Other tests are then ordered, depending on the clinical situation.

Treatment is aimed at preventing recurrence (by treating atherosclerotic risk factors and possible causes of embolism) and avoiding progression to a full-blown attack. Anti-platelet agents are used, and sometimes surgery or percutaneous transluminal angioplasty. Close, long-term monitoring is essential, as the risk of recurrence or stroke is relatively high.

NOTE.–

In everyday language, the brain (or encephalon) comprises two parts called *cerebral hemispheres*. Each hemisphere is divided into four areas called *lobes*, in which these different functions are managed. The *frontal lobes*: speech and language, reasoning, memory, decision-making, personality, judgment, movement. The right frontal lobe handles movements on the left side of the body, while the left frontal lobe handles movements on the right. The other lobes are the *parietal lobes*: reading, spatial orientation, sensitivity. Here too, the right parietal lobe manages sensitivity on the left side of the body and vice versa; the *occipital lobes*: vision, the *temporal lobes*: language, memory, emotions. Therefore, on the basis of symptoms, *the doctor can suspect a stroke and guide the diagnostic examination.* For example, sudden *weakness* of an arm or *leg on the left side may be due to a defect in the vascularization of a cerebral artery* located in the right frontal lobe regions, which manage movements on the left side of the body.

VOCABULARY CORNER.–

– *Arteriography*: medical imaging test used to study the morphology of arteries. It involves observation of the vessels and their functioning using X-rays. The radiologist places a small tube called a catheter into the artery of interest, injecting an iodinated contrast medium to make the arteries visible. The examination may be for diagnostic or therapeutic purposes.

– *Diffusion-weighted MRI*: magnetic resonance imaging technique used to measure the microscopic displacements of water molecules in biological tissues.

– *Hematoma*: an effusion (*abnormal flow*) of blood under the skin. It occurs when the small blood vessels under the skin are damaged and blood collects in the space between the skin and muscles.

– *Hypoperfusion*: reduction in blood flow to an area of the body (to be distinguished from "hypovolemia": reduction in blood volume).

3.8.1.1.4. Stroke figures worldwide

Strokes: figures by age [BÊM 19]

The risk of stroke increases with age. Nearly three-quarters of patients are over 65. After the age of 50, the incidence doubles every 10 years. According to the France AVC association, the number of strokes per year is estimated as a function of age:

Age	Estimated number of strokes per year per 100,000 inhabitants
15–45 years	10–30
5–64 years	170–360
65-74 years	490–890
Age 75 and over	1,350–1,790

Table 3.9. *Estimated number of strokes per year*

Stroke after-effects in figures

– A quarter of sufferers die within a relatively short time.

– Half survive but are left with varying degrees of disability.

– A quarter survive without any after-effects.

– Stroke is the first cause of acquired disability in adults.

– It is the second leading cause of dementia.

– It is the second leading cause of death in men, first in women.

– 30% of stroke victims will suffer from depression.

The mortality rate at one month is 20%, and 32–60% at three years. In addition, 300,000 former stroke victims are now disabled. Lastly, stroke is a major financial burden. In the UK, it is estimated to represent 5% of the annual health budget.

Risk factors in figures

Many strokes could be prevented by reducing certain risk factors. Ten risk factors are associated with 90% of strokes:

1) *hypertension* is implicated in 40% of strokes, multiplying the risk by a factor of 2, and by a factor of 5 in people under 55;

2) smoking: the risk of stroke is multiplied by 3 for smokers;

3) *abdominal obesity*: involved in 36% of strokes;

4) unbalanced diet: associated with 33% of strokes;

5) lack of physical activity;

6) alcohol consumption;

7) *atrial fibrillation*: the risk is multiplied by 4, making it the leading risk factor for heart disease;

8) psycho-social factors: stress, depression and social isolation are significant risk factors;

9) diabetes: is involved in ischemic strokes;

10) high levels of certain lipids in the blood (cholesterol, triglycerides, etc.).

VOCABULARY CORNER.–

– *Arterial hypertension* (AH): defined as excessive pressure in the arteries, which persists over time.

– *Atrial fibrillation (AF)*: is characterized by an irregular and often very rapid heart rate. This is an arrhythmia problem. Arrhythmia is an abnormality that affects the normal heart rate. In the presence of arrhythmia, the heart tends to beat too slowly (bradycardia), too fast (tachycardia) or irregularly. This arrhythmia can cause symptoms such as heart palpitations, fatigue or shortness of breath.

NOTE.–

Blood pressure, also known as arterial pressure, is the pressure exerted by blood on the arteries as it circulates through the body. It is expressed in millimeters of mercury (mmHg). Tensioning the arterial walls enables the blood to circulate properly and reach, via the arteries and capillaries, the furthest and tiniest parts of our body, from our scalp to our toes. Capillaries are the very small (*hair-thin*) vessels through which blood exchanges with tissues. Blood pressure is made up of two values: the first is *systolic pressure*, corresponding to the pressure of blood ejected by the heart into the circulatory system. This is the maximum pressure during heart muscle contraction. It lies between 90 and 130 mmHg, or even 140 mmHg. The second value is *diastolic pressure*, corresponding to the pressure on the arteries as the heart fills with blood. This is the minimum pressure when the heart muscle is at rest. It lies between 60 and 90 mmHg. Both values depend on gender and age. For example: systolic *pressure* increases with *age* for both *sexes*, but more rapidly in women, catching up with men at around age 60, at 130 mmHg. Thereafter, it increases more in women, reaching an average of 145 mmHg at age 80, compared with 140 mmHg in men. Furthermore, below the threshold of 90 mmHg systolic pressure and 60 mmHg diastolic pressure, the individual is *hypotensive*. While hypotension is not a disease, and is not fatal, it can significantly impair the quality of life, and can be a symptom of pathology.

3.8.1.1.5. Key facts about stroke risk factors

The main risk factors for stroke are smoking, obesity, diabetes and atrial fibrillation.

Smoking

According to the WHO [WOR 23b], the main facts about tobacco as of 2020 are as follows:

– Tobacco kills up to half of those who use it.

– Tobacco kills more than 8 million people every year. More than 7 million of these are current or former users, and around 1.2 million are non-smokers involuntarily exposed to smoke.

– More than 80% of the world's 1.3 billion smokers live in low- and middle-income countries.

Abdominal obesity

Abdominal obesity is excess weight that develops over time around the center of the body, also known as visceral fat (visceral fat is characterized by the accumulation of fatty deposits between the abdominal muscles and organs). Stress and high sugar intake can contribute to its appearance [ARG 20].

According to the WHO [WOR 23b], the main facts about obesity are as follows:

– Worldwide, the number of cases of obesity has almost tripled since 1975.

– In 2016, more than 1.9 billion adults – people aged 18 years and over – were overweight. Of these, more than 650 million were obese.

– 39% of adults aged 18 years and over were overweight in 2016, and 13% were obese. Most of the world's population lives in countries where overweight and obesity cause more deaths than underweight (*weight below normal weight for height*).

– Most of the world's population lives in countries where overweight and obesity kill more people than underweight.

– In 2016, more than 340 million children and adolescents aged 5–19 years were overweight or obese.

– In 2019, 38 million children under the age of 5 were overweight or obese.

– Obesity can be prevented.

Diabetes

Diabetes is a chronic disease that occurs when the pancreas does not produce enough *insulin*, or when the body does not properly use the insulin it does produce. Insulin (see note) is a *hormone* that regulates the concentration of sugar in the blood. *Hyperglycemia*, or high blood sugar levels, is a frequent effect of uncontrolled diabetes. Over time, it leads to serious damage to many organ systems, particularly nerves and blood vessels.

There are two types of diabetes:

Type 1 diabetes: formerly known as insulin-dependent or juvenile diabetes, it is characterized by insufficient insulin production, which must be administered daily. The cause of type 1 diabetes is unknown, and as far as we know, it cannot be prevented.

Type 2 diabetes: formerly known as non-insulin-dependent or mature-onset diabetes, it results from the body's poor use of insulin. Type 2 diabetes accounts for the majority of diabetes cases. It is largely the result of excess weight and lack of physical activity. Its symptoms may be similar to those of type 1 diabetes, but are often less marked. As a result, the disease can be diagnosed several years after onset, when complications are already present [WOR 23b].

Similarly, according to the WHO [WOR 23b], the main facts about diabetes are as follows:

– The number of people with diabetes rose from 108 million in 1980 to 422 million in 2014.

– Diabetes is a major cause of blindness, kidney failure, heart attacks, strokes and lower limb amputations.

– Between 2000 and 2016, premature mortality attributable to diabetes increased by 5%.

– An estimated 1.6 million deaths were directly caused by diabetes in 2016, while 2.2 million deaths were attributable to hyperglycemia in 2012.

– A healthy diet, regular physical activity, a normal weight and avoiding tobacco consumption are all ways of preventing or delaying the onset of type 2 diabetes.

– A healthy diet, physical activity, medication, regular screening and treatment of complications help to treat diabetes and avoid or delay its consequences.

NOTE.–

i) Insulin, with empirical formula $C_{257}H_{383}N_{65}O_{77}S_6$, is a protein hormone secreted by the *β*-cells of the islets of Langerhans in the pancreas. In a non-diabetic person, insulin is secreted continuously and regulates blood glucose levels.

ii) Islets were described in 1868 by the German biologist Paul Langerhans (1847-1888). They were named "islets of Langerhans" in his honor. Islets of Langerhans account for around 1-2% of pancreas mass, or 1-1.5g. An adult has between 750,000 and 1,500,000 islets, most of them located at the back of the pancreas. Ranging in size from 50 to 900 μm, islets are made up of a few cells in the case of the smallest islets, and around 5,000 cells in the case of the largest.

3.8.2. *Thyroid and parathyroid scans*

As explained in section 3.4, in thyroid scintigraphy, radioactive iodine-123 selectively binds to the thyroid to detect morphological or functional abnormalities. Similarly, we have seen that in radiotherapy, iodine-131 in the form of sodium iodide $Na^{131}I$ is indicated for the treatment of thyroid cancer. It is therefore necessary to understand the purpose and function of the thyroid, as well as the mechanism of thyroid hormone secretion, which determines the various forms of *hyperthyroidism* (*Graves' disease, basedowified goiter, hot nodules, toxic nodules, etc.*). Hyperthyroidism is defined as all disorders resulting from excessive stimulation of their target tissues by the thyroid hormones *triiodothyronine* (T_3) and *tetraiodothyronine* (T_4) or *thyroxine* [WAQ 17]. The structures of T3 and T4 are shown in Figure 3.11.

3.8.2.1. *General information*

3.8.2.1.1. Structure and importance of the thyroid gland

The *thyroid gland* is located in the anterior cervical region, on top of the trachea, in front of the vocal cords and larynx. It is located on the anterior surface of the base of the neck (Figure 3.27).

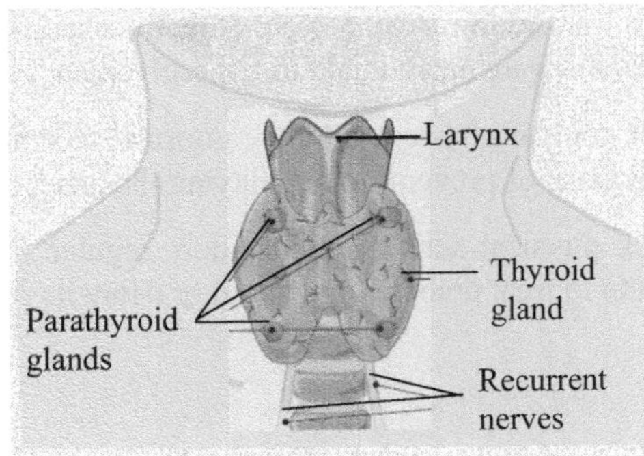

Figure 3.27. *Diagram showing the location of the thyroid gland. The recurrent nerves passing through the thorax control the movement of the larynx and vocal cords. For a color version of this figure, see www.iste.co.uk/sakho/nuclearphysics3.zip*

The weight of the normal thyroid gland increases from around 1.5 to 20 grams from birth to adulthood. The volume of the normal thyroid gland is 7 to 10 ml. These dimensions also increase with the individual's corpulence. In addition, the normal thyroid gland is larger in men than in women. It should be noted that these dimensions vary with iodine intake. Indeed, the size of the normal thyroid gland increases with deficiency to capture more available iodine, and decreases with iodine intake.

Lymph nodes are small, rounded organs scattered throughout the body along the path of lymph *circulation*. Lymph is a biological fluid containing numerous cells involved in the body's defense system, notably leukocytes or white blood cells. The main function of the lymph nodes is to *clean the lymph* and rapidly trigger *defense reactions in the event of the arrival of external agents such as viruses or bacteria.* They therefore play a *key role in the immune system* [GIO 20].

3.8.2.1.2. Parathyroid glands

Classically numbering 4 (see Figure 3.27), the *parathyroid glands* are located behind the thyroid gland and regulate calcium levels in the blood via the hormone *parathyroid hormone* (PTH). In some cases, one of these glands may become hyperfunctional, secreting far too much PTH. *Parathyroid scintigraphy* can detect hyperactivity in one or more of these glands. The parathyroid glands are part of the *endocrine system*, a collection of glands and cells that produce hormones and release them into the bloodstream to regulate many bodily functions, such as breathing and blood circulation [DES 15, GIO 20].

Most calcium is stored in the bones. When blood calcium levels are low, the parathyroid glands produce PTH so that the bones release calcium into the blood.

Conversely, when blood calcium levels are high, the parathyroid glands produce less PTH.

Sometimes, the cells of any of the parathyroid glands undergo changes that make their growth pattern or behavior abnormal. These changes can lead to the formation of non-cancerous (benign) tumors called *parathyroid adenomas*. They can also lead to a non-cancerous condition called *parathyroid hyperplasia*. In some cases, changes in the cells of the parathyroid glands can cause *parathyroid cancer* (also called *parathyroid carcinoma*). This is a very rare cancerous tumor [CIM 17].

NOTE.–

Calcemia is the level of calcium in the blood. Almost all the calcium in the body is stored in the bones. The amount of calcium in the blood is strictly regulated: if calcium levels fall (known as *hypocalcemia*: total plasma calcium concentration < 8.8 mg/dL (< 2.20 mmol/L)), the bones release some into the bloodstream to restore the balance. However, if the level is too high (known as *hypercalcemia*: total plasma calcium concentration > 10.4 mg/dL (> 2.60 mmol/L) or ionized calcium > 5.2 mg/dL (> 1.30 mmol/L)), the extra calcium is stored in the bones or eliminated in the urine or feces. The level of *ionized calcium* (free) in the blood is normally between 4.6 and 5.3 mg/dL. It should be noted that the so-called normal values for calcium in the blood may vary, depending on the laboratory where the tests are carried out. High blood calcium levels may be a sign of *hyperparathyroidism* (abnormally high production of parathyroid hormones by the large parathyroids, leading to an increase in blood calcium levels), cancer, etc.

VOCABULARY CORNER.–

– *Endocrinology*: medical specialty concerned with hormones and their effects on how the body functions.

– *Endocrine gland*: an internal gland that secretes hormones into the bloodstream directly, rather than via a duct like an exocrine gland. Endocrine glands are found in most animals, including invertebrates.

3.8.2.2. *Thyroid hormones*

3.8.2.2.1. Structure and synthesis

There are three *thyroid hormones* (TH): *triiodothyronine* or T_3 *tetraiodothyronine* (thyroxine) or T_4 and *thyrocalcitonine* (see note). The numbers 3 and 4 in T_3 and T_4 correspond to the number of iodine atoms required to synthesize the hormone. *Follicular cells* produce thyroid hormones in the form of T_3 and T_4, which have a common organic structure: *thyronine*. This molecule is derived from the amino acid tyrosine. T_3 is therefore 3,5,3-tri-iodo-L-thyronine and T_4 is 3,5,3',5'-tetra-iodo-thyronine (Figure 3.28).

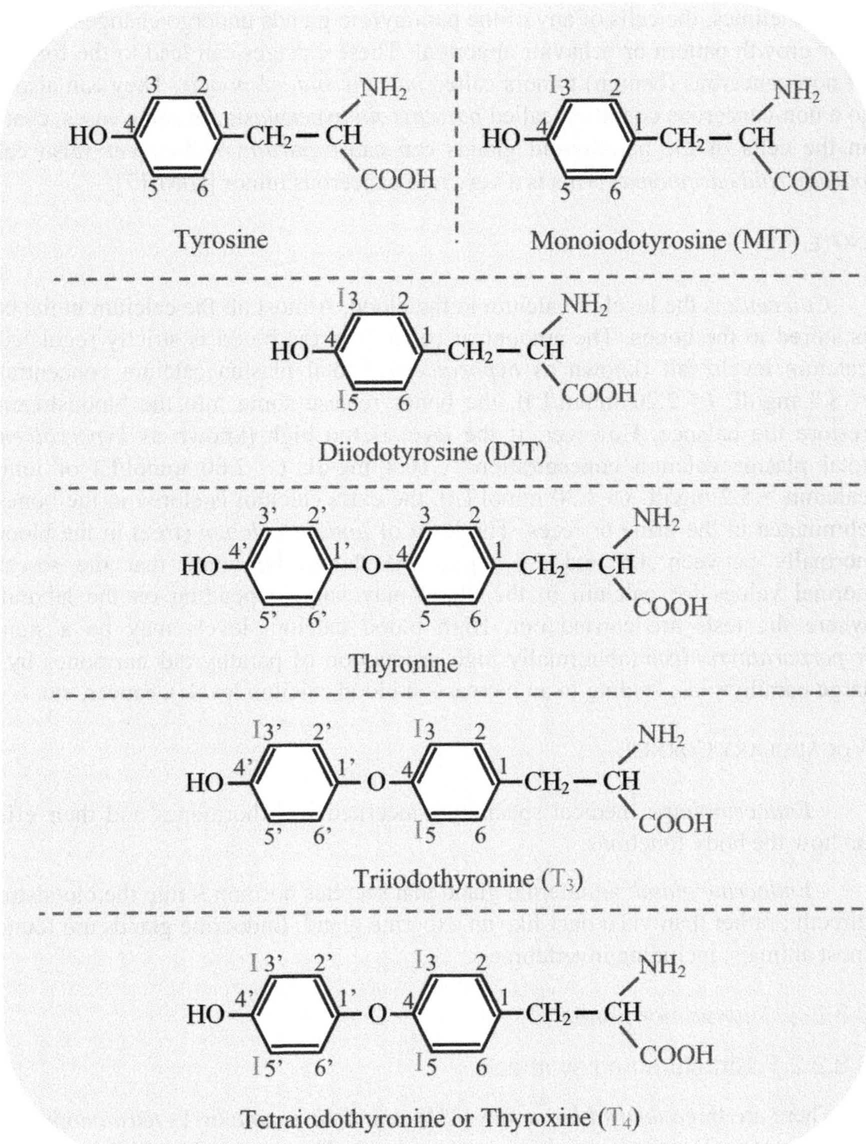

Figure 3.28. *Structures of the thyroid hormones T_3 and T_4 with the same common structure: thyronine. For a color version of this figure, see www.iste.co.uk/sakho/nuclearphysics3.zip*

Thyroid hormones are involved in *metabolism* (*a set of biochemical reactions that enable the body to maintain a state of equilibrium*). They also play a part in brain development and enable the respiratory, cardiac and nervous systems to function optimally. These hormones also regulate body temperature, muscle tone, menstrual cycles, weight and cholesterol levels. In this sense, hyperthyroidism leads to dysfunctions, of varying degrees, in these different bodily functions [WAQ 17]. Thyroid hormones are synthesized from iodine I atoms and an amino acid: tyrosine (Figure 3.28) under the action of an enzyme *thyroperoxydase* (TPO).

Once synthesized, hormones are either stored inside the thyroid gland in the colloid or released into the bloodstream. The thyroid mainly produces thyroxine (T_4), which is then converted into triiodothyronine (T_3). Triiodothyronine is the active hormone, mainly in the liver and skeletal muscle [TOU 18].

Thyroid hormones are prescribed to treat hypothyroidism (low levels of hormones in the blood) or thyroidectomy (surgical removal of the thyroid gland). The hormones used are triiodothyronine (T_3) and tetraiodothyronine (T_4). T_3 has a half-life of 24 h, so its use would require two to three doses a day; whereas, T_4, which has a half-life of 6–8 days, can be taken once a day. This justifies the preferential use of T_4 (although T_3 is more effective at the receptor level, since T_4 is converted into T_3 by peripheral tissues; treatment with T_4 allows normal T_3 levels to be achieved).

NOTE.–

– *Thyroperoxydase (TPO)*, also known as *thyroid peroxidase*, is an oxidoreductase which catalyzes the first stage in the biosynthesis of thyroid hormones from tyrosine. Its activity is increased by thyroid stimulating hormone (TSH).

– *Calcitonin or thyrocalcitonin* is a hormone secreted by the thyroid cells. Its role is to regulate phosphorus and calcium metabolism: it reduces calcium fixation in the bone and increases urinary elimination of calcium and phosphorus. Calcitonin is produced in excess in certain diseases, including medullary (*central*) thyroid cancer. Measuring calcitonin helps to diagnose this cancer, as well as makes it possible to assess the effectiveness of treatment or the occurrence of relapse [BÊM 18].

3.8.2.2.2. Role and benefits of iodine

The only known role of iodine in humans is that it is the essential element required for the synthesis of the thyroid hormones thyroxine (T_4) and triiodothyronine (T_3).

Iodine is a trace element present in small quantities in adult humans, with 10 to 20 milligrams concentrated in the thyroid gland. The daily iodine intake recommended by the WHO is 150 μg/day, and adequate intake is mainly provided by *iodized salt* and seafood, particularly seaweed, crustaceans and marine fish. Seawater is in fact the main source of iodine: up to 5 parts per million (mg/L). Iodine-enriched table salt is the simplest and most effective food source for increasing iodine intake in iodine-deficient regions.

In the body, iodine is distributed in a diffusion space of around 35% of body weight, including the thyroid (organic iodine), salivary, gastric and mammary glands, the vascular sector and elimination organs (inorganic iodine). Iodine is present in feces and sweat, but the kidney is the main emunctory (*elimination organ*): daily clearance of around 30 mL/min [THO 18].

Iodine is metabolized in the body through a multitude of processes involving the hypothalamus, pituitary gland, thyroid gland and blood. Consumed iodine enters the bloodstream in the form of plasma inorganic iodide. The thyroid uses iodide to *synthesize thyroid hormones*. Excess iodide is eliminated in the urine [LEM 23]. Iodide is an essential element in the biosynthesis of T_3 and T_4 hormones, which are essential for the development and maturation of the central nervous system, lungs and skeletal muscle in fetuses and newborns [NAD 14].

In addition, iodine deficiency may be manifested by the appearance of a number of typical symptoms: fatigue, dry skin, increased sensitivity to cold, moodiness and weight gain. When the deficiency is severe, these symptoms may be followed by *enlargement of the thyroid gland (goiter)*, which may be so large as to impede breathing [LEM 23].

NOTE.–

Iodized salt is food salt enriched with iodine to reduce the risk of iodine deficiency. Iodization is the enrichment of dietary salt with iodine, which must contain between 15 and 20 mg of iodine per kg of salt.

3.8.2.2.3. Thyroid hormone secretion mechanism

The pituitary gland, which secretes TSH, is the real controller of the thyroid. It is an *endocrine gland* located in a bony cavity (the pituitary fossa) and connected to the hypothalamus in the brain (Figure 3.29).

Thyroid hormone secretion is stimulated by the pituitary hormone thyroid stimulating hormone (TSH). TSH secretion is itself stimulated by a hormone from the hypothalamus, thyrotropin releasing hormone (TRH). TSH controls the secretion of T_3 and T_4 hormones. When thyroid hormone levels in the blood are too low (hypothyroidism), the pituitary gland releases more TSH. This stimulates the thyroid to produce more T_3 and T_4. Conversely, if thyroid hormone levels are too high (hyperthyroidism), the endocrine gland in the brain responds by decreasing its release of TSH. This slows the production of T_3 and T_4 [DES 15, WAQ 17].

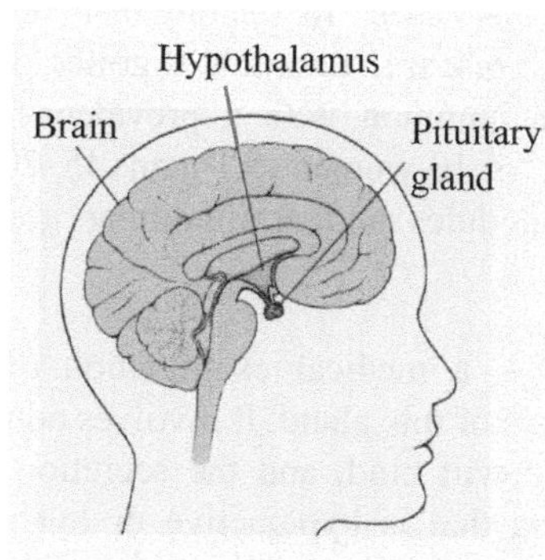

Figure 3.29. *Location of the pituitary gland at the base of the brain. For a color version of this figure, see www.iste.co.uk/sakho/nuclearphysics3.zip*

The production and release of the thyroid hormone in the thyroid gland is regulated by a feedback system in the brain – the hypothalamus and pituitary gland – which produce TRH and TSH, respectively. If everything works as planned, the body makes what it needs and T_4 is converted into T_3. If T_3 production is insufficient, or if the T_4 produced is not properly converted into this active thyroid hormone, the whole system goes haywire. The metabolism and mitochondria (*the little factories in the cells that convert food and oxygen into energy*) do not get the right signals, and weight gain and symptoms of hypothyroidism occur. In addition, this can lead to chronic inflammation, unbalanced insulin levels and impaired blood sugar metabolism, all of which compromise health and the ability to lose weight [MAR 21].

In pregnancy, hyperthyroidism can have more serious consequences for both mother and child. It can lead to spontaneous abortion, premature delivery, malformations in the fetus or functional disorders in the child. Close monitoring of pregnant women suffering from this condition is therefore essential [WAQ 17].

NOTE.–

Thyroxine (T_4) was isolated in 1910 by the American chemist Edward Calvin Kendall (1886–1972) from 3 tons of pig thyroid gland. As for triiodothyronine (T_3), it was discovered in 1952 by the French physician Jean Roche (1901–1992) following his work on biochemistry and its physiological and medical aspects, the best known of which were recognized as being on thyroid hormones, respiratory pigments, proteins, enzymes and ossification.

3.8.2.2.4. Principle of thyroid scintigraphy

Thyroid scintigraphy is used to distinguish between different forms of hyperthyroidism (Graves' disease, basedow's goiter, hot nodules, toxic nodules, etc.). Hyperthyroidism is common with a prevalence of 0.5–2% in the general population and a sex ratio of 10 women to 1 man. 45–70% of cases are the result of Graves' disease. Toxic nodules and multinodular goiter predominate in elderly subjects [WAQ 17].

Thyroid scintigraphy is a medical examination of the thyroid gland, which allows a functional analysis of this gland. It involves administering slightly modified iodine, which the thyroid will bind, and the secretion of labeled iodine makes it possible to locate an area that is hyperactive or hypo-active. If the substance is evenly distributed, this means that the thyroid is functioning correctly. Thyroid scintigraphy is combined with biological tests, including TSH levels and thyroid ultrasound [DEL 19].

Parathyroid scintigraphy is used to study the functioning of the parathyroid glands, using radioactive products injected in small quantities. It is used to study the origin of certain parathyroid function disorders (adenoma, hyperplasia, etc.).

3.8.2.2.5. Graves' disease: causes and symptoms

Graves' disease is an autoimmune hyperthyroidism (thyroid disease). It is caused by a deficiency in the immune system in which the *TSH* (*T*hyroid-*S*timulating *H*ormone) receptors are the target of autoantibodies, leading to excessive synthesis of thyroid hormones. These antibodies are called anti-TSH receptors, otherwise known as: *TRAK* (*T*SH-*R*ezeptor-*A*utoanti*K*örper). By binding to TSH receptors, autoantibodies trigger the secretion of thyroid hormones because their structure is close to that of TSH [DES 15]. In this way, the person affected produces abnormal antibodies directed against the follicular cells of the thyroid. Follicular cells produce thyroid hormones in the form of T_3 and T_4.

Graves' disease is diagnosed when the TRAK antibody test is positive. The therapeutic treatment of this disease depends directly on the level of TRAK antibodies measured in the blood. Other antibodies may also be involved in the development of Graves' disease. These affect between 30% and 50% of patients. This disease can affect any individual. Young women aged between 20 and 30 are more likely to be affected by the disease [WAQ 17].

3.8.2.2.6. Thyroid nodules. Basedow's goiter

By definition, a *nodule* is any localized enlargement of the thyroid gland (*nodulus* = small node) [WEM 11]. Generally, thyroid nodules are hypofixing, i.e. they do not fix iodine. There are two types of nodule depending on whether or not they produce thyroid hormones:

– Cold nodules on scintigraphy: these nodules cause no change in thyroid hormone secretion and therefore no directly related signs. These nodules are also known as *hypofixing nodule*.

– Hot nodules on scintigraphy: these nodules secrete excess thyroid hormones. These so-called toxic autonomous nodules are also known as *hyperfixating nodules*.

In addition, goiter is the name given to the increase in volume (hypertrophy) of the thyroid gland. In plain English, the appearance of a goiter shows that the thyroid gland is larger than normal (Figure 3.30).

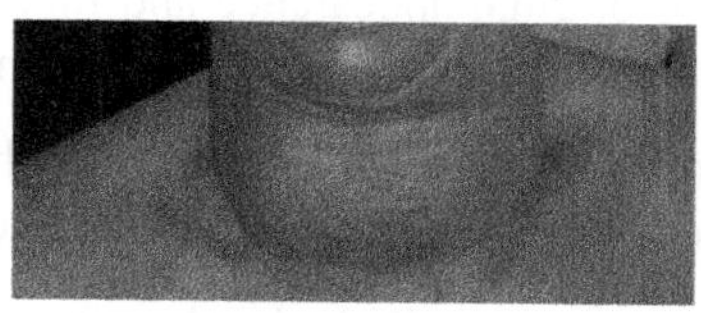

Figure 3.30. *Photograph of a goiter [DES 15]. For a color version of this figure, see www.iste.co.uk/sakho/nuclearphysics3.zip*

A *non-toxic goiter* is one in which the thyroid functions normally, without being inflammatory or tumorous in origin. A *toxic goiter* is a goiter with hyperthyroidism, for example, in Graves' disease. In Graves' disease, the goiter is painless. In plain English, diffuse, homogeneous (involving the entire thyroid gland) and vascular (i.e. without nodules). This is known as *Grave's disease*. Graves' disease can also occur on a pre-existing goiter, in which case it is known as *Basedow's goiter*. If there is any doubt about the diagnosis, an ultrasound can be used to determine the size of the thyroid: a goiter is considered to exist if the estimated mass is greater than 20g [DES 15].

NOTE.–

Robert James Graves (1797–1853) was an Irish physician and surgeon. In medicine, he is credited with the first description of Graves' disease. However, he believed that it was heart palpitations that caused the thyroid gland to enlarge. In 1840, the German physician Carl Adolph von Basedow (1799–1854) gave a correct description of the disease. This is how this thyroid condition came to be known as Graves–Basedow disease.

3.8.2.3. *Thyroid cancer*

3.8.2.3.1. General information

Thyroid cancer is one of the disorders affecting the thyroid gland. This cancer is rare, accounting for 1% of all cancers. Thyroid cancer is mainly due to exposure to radiation, iodine deficiency or excessive iodine intake. There are two possible scenarios in endocrinology: either the cancer is strictly localized in the gland, or it is a cancer with extension and lymph node involvement [BUI 19].

In general, thyroid cancer is a cancer that affects one of the different cell types that make up the thyroid. Although it is one of the main endocrine cancers, 95% of thyroid nodules turn out to be benign. Of the remaining 5%, which are malignant tumors, there are several histological types of thyroid cancer:

– Differentiated carcinomas (90%): papillary (the most common, accounting for over 80% of reported thyroid cancers) and vesicular (or follicular), which develop from cells that retain their original appearance and functions (production of thyroid hormones and iodine fixation), are not very aggressive and progress slowly.

– Undifferentiated or anaplastic carcinomas, in which the cells have lost most of their functionality, are very rare and are mostly found in elderly subjects.

– Rare medullary carcinomas, which develop from specific thyroid cells, the parafollicular cells or C cells.

Source: https://en.wikipedia.org/wiki/Thyroid_cancer.

Note that the parafollicular cells, or C cells, are thyroid cells that produce and secrete calcitonin. C cells are paler in color than thyroid cells. Calcitonin or thyrocalcitonin is a hormone whose role is to regulate the metabolism of phosphorus and calcium: it reduces calcium fixation in the bone and increases urinary elimination of calcium and phosphorus. Its measurement contributes to the diagnosis of thyroid cancer, as well as making it possible to judge the effectiveness of its treatment or the occurrence of relapse [BÊM 18].

VOCABULARY CORNER.–

– *Endocrine system*: all the organs that have the capacity to release hormones into the bloodstream.

– *Malignant tumor*: a cluster of cancerous cells. A benign tumor is not cancer, but a malignant tumor is.

– *Histology*: A branch of biology that deals with the structure of living tissues. In oncology, it is used to determine whether a *tumor* is malignant or not.

The main sign of thyroid cancer is the presence of a nodule at the base of the neck. This nodule is generally painless, not bothersome and very hard, and is accompanied by inflammation of the lymph nodes in the neck (Figure 3.31).

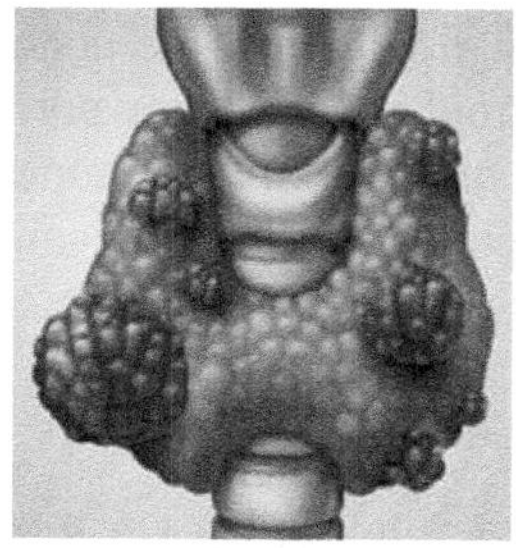

Figure 3.31. *Thyroid nodules at the base of the neck due to thyroid cancer [BUI 19]. For a color version of this figure, see www.iste.co.uk/sakho/nuclearphysics3.zip*

In some cases, thyroid cancer can alter the voice, making it hoarse. It may also be responsible for the formation of a goiter. In practice, papillary cancers easily spread to the lymph nodes, but rarely lead to the formation of metastases. Follicular cancers mainly lead to bone and lung metastases [ASS 21].

3.8.2.3.2. Diagnosis

In the vast majority of cases, thyroid cancer is diagnosed by cervical ultrasound, cytopuncture of the nodule and anapathological examination of the cells. A CT scan, MRI or thyroid scintigraphy may be performed in certain cases [BUI 19].

Furthermore, when thyroid cancer is discovered, the tumor cells may subsequently spread to more distant lymph nodes, leading to the formation of metastases. Once a nodule has been discovered, a full work-up is required to determine whether the thyroid abnormality is malignant (cancerous) or benign (non-cancerous). In 95% of cases, the nodules detected are benign: adenomas or cysts. The diagnostic work-up always begins with a full clinical examination, including palpation of the thyroid and a medical questionnaire about personal and family history of thyroid disease. This is followed by a number of tests [ASS 21]:

VOCABULARY CORNER.–

Anapathological examination: examination that consists of studying tissues or cells taken from a patient to identify and analyze abnormalities linked to a disease. The examination is carried out first with the naked eye and then under a microscope by a doctor called an anatomopathologist or pathologist.

Thyroid assessment

Thyroid tests are routinely requested by the doctor. This consists of a series of biological tests carried out on the basis of a blood sample. In general, it does not enable the malignant or benign nature of the nodule to be determined, but it does reveal whether or not the thyroid is functioning normally. In most cases, only TSH is measured and the concentration of thyroid hormones (T_3, T_4) is only requested when there are clinical signs of hypo- or hyperthyroidism. Calcitonin and calcium levels in the blood are sometimes requested preoperatively.

Cervical and thyroid ultrasound

An ultrasound examination of the neck area is routinely performed. It is a painless examination that allows the structure of the internal organs to be observed. During the scan, the doctor measures the dimensions of the thyroid gland, the number, size and shape of nodules and detects the presence of any other abnormalities. The doctor also looks to see if the neighboring lymph nodes are normal. Above all, they look for clues that may indicate cancer: the first of these is the way in which the nodule is perceived on ultrasound. Quite frequently, the nodules observed appear white (they are said to be hyperechoic); more rarely, they appear black (anechoic). In both cases, the nodule is almost always benign. Isoechogenic nodules (the same color as the rest of the thyroid) or dark gray nodules (hypoechoic) may be benign or malignant. Finally, the appearance of the lymph nodes adjacent to the thyroid gland is examined for any abnormalities.

Cytological puncture

Cytology is the part of biology that studies living cells. Cytological puncture involves taking samples of cells.

The ultrasound results make it possible to determine whether it is necessary to puncture cells within nodules to observe them under the microscope. The doctor will suggest puncturing a nodule (usually more than 1 cm long) if there are associated risk factors that increase the likelihood of malignancy. These include a history of external radiotherapy in childhood, a genetic predisposition or a family disease that puts the patient at risk of thyroid cancer, persistently high levels of calcitonin, abnormal lymph nodes, and nodules that have recently grown or look risky.

In practice, the puncture is performed at the same time as the ultrasound, which guides the doctor's action. It is carried out without anesthetic because the needle is so thin that it is not painful. The cells are examined under a microscope to determine whether the nodule is benign, suspect or malignant. Sometimes, the sample cannot be interpreted. In this case, the doctor will suggest a new puncture or the removal of the suspect nodule. When the nodule measures less than one centimeter, simple monitoring is often recommended, as the risk of cancer is low. A new ultrasound scan is scheduled within 6 to 18 months to check that the nodule is benign.

Additional examinations

Some tests are not systematically carried out, but are offered to certain patients, depending on the characteristics of their disease:

– Thyroid scintigraphy may be indicated when the biological work-up shows a low TSH level. This examination, which maps thyroid activity using a radioactive product (technetium or iodine-123), can identify a hyperfunctional nodule, i.e. one that produces excess thyroid hormones.

– A CT scan of the neck (or cervical scanner) and an MRI (magnetic resonance imaging) are imaging tests prescribed when ultrasound has not provided a sufficiently precise description of the abnormalities identified in the thyroid gland.

Assessment of extension

Extension tests are used to determine whether thyroid cancer cells have spread to neighboring organs (mainly lymph nodes) or to distant organs. To do this, several additional tests may be carried out:

– *The ENT examination* checks the condition and mobility of the vocal cords located just above the thyroid gland.

– *CT scan or MRI*: a CT scan of the neck (or cervical scan) and an MRI (magnetic resonance imaging) may be prescribed to assess whether the thyroid cancer has spread to neighboring lymph nodes.

– *PET scan*: this is an imaging test that combines a scintigraphy with a weakly radioactive marker. It shows the metabolic activity of thyroid cells. It is mainly used to assess the spread of *anaplastic cancers*.

The prognosis for thyroid cancer is relatively good, with over 90% survival at 5 years [BUI 19]. However, it should be borne in mind that this prognosis depends on several parameters [ASS 21]:

– In adults, the risk of relapse increases with the extent of the tumor and the patient's age. In children, relapses are relatively frequent.

– Life expectancy depends mainly on age and the formation of metastases. For example, the risk of death is low in children and higher in the over 65 years of age.

3.8.2.3.3. Treatment

Without treatment for thyroid cancer, thyroid tumors grow and can invade the lymph nodes adjacent to the thyroid [ASS 21].

Thyroid cancer is treated by thyroidectomy, which involves removing the gland. Depending on how far the cancer has spread, the patient may then be treated with iratherapy. The patient will take a radioactive iodine tablet orally, which will attach itself to the remaining thyroid tissue and destroy it. The patient will then be under medical supervision for several years [BUI 19]. The reason for taking iodine tablets is as follows [BUI 19].

Taking a non-radioactive iodine tablet as a preventive measure saturates the thyroid gland. In this case, the thyroid gland is no longer able to bind the radioactive iodine inhaled or ingested during the contamination. The radioactive iodine is thus eliminated in the urine, protecting against the development of thyroid cancer.

The iodine tablet, in the form of potassium iodide, is taken as a single dose of 130 mg. This tablet can be taken again if necessary over the following days. Every person living near a nuclear facility must have iodine tablets available.

The dosage for taking non-radioactive iodine tablets dissolved in a glass of water as a preventive measure is as follows:

– one tablet for adults, pregnant women and children aged over 12 years;

– half a tablet for children aged 3–12 years;

– a quarter of a tablet for children aged 0–3 years.

VOCABULARY CORNER.–

– *Anaplastic thyroid cancers*: undifferentiated malignant tumors developed at the expense of the thyroid parenchyma. Anaplastic cancers account for around 1.6% of thyroid cancers and have an annual incidence of around two per million population [WEM 08].

– *Parenchyma*: tissue of an organ that ensures its function.

– *Thyroid parenchyma*: this is made up of more or less spherical thyroid follicles, which are often deformed because they are packed together, and vary in size, depending on their state of secretory activity. The thyroid gland contains numerous small rounded sacs called follicles. The follicles produce, store and release thyroid hormones. The thyroid is made up of different types of cells. The follicles are lined with follicular cells.

– *Iratherapy:* treatment using radioactive iodine (Iodine-131). This has been used for several decades to treat certain forms of hyperthyroidism (Graves' disease, hot nodule, toxic multinodular goiter). It is an effective alternative to drugs (synthetic antithyroid drugs) and surgery.

NOTE.– [DIR 99]

The first option for treating thyroid cancer is surgery. However, it is often impossible to remove all the cancerous tissue and metastases can develop. Therefore, even after surgery, treatment with radioactive iodine is usually applied to destroy the remaining cancer cells. Thyroid cancer cells partially lose their ability to bind to iodine, so iodine retention in thyroid cancer cells is lower than in normal thyroid tissue. Often, after surgical removal of thyroid cancer tissue, an initial activity of around 3,000 MBq is administered to eliminate thyroid remnants. A much higher dose of up to 8,000 MBq of iodine-131 is then required to treat any metastases. When no normal tissue remains after thyroidectomy, but metastases exist, high levels of radioactivity are administered immediately. Repeated treatment may be necessary in both cases. Since only a small amount of thyroid tissue remains after surgical treatment and the binding process is disrupted, only a small proportion of the activity administered will be bound. When thyroid or tumor tissue does not concentrate iodine, around 80% of the activity administered will usually be excreted within 48 h. If 5,000 MBq are administered to treat thyroid carcinoma, residual activity in the patient will be less than 1,000 MBq in 48 h and 750 MBq in 72 h. Excretion is much longer in subjects with numerous cancerous metastases.

Finally, we should mention the impact of the nuclear accident at Chernobyl in 1986 on children's thyroids.

Radioactive iodine released into the environment can be inhaled or ingested by people exposed to it. It passes into the bloodstream and attaches itself to the thyroid gland, the target organ for iodine. This increases the risk of thyroid cancer for those exposed, particularly children, which can occur years after irradiation. Let us briefly summarize the facts reported in the literature after the Chernobyl accident.

On April 26, 1986, at around 1 a.m., the RBMK (boiling water high power reactor) No. 4 at the Chernobyl power station in Ukraine was destroyed. In the days following the accident, 17,000 children received a sievert (1 Sv), a dose well above the permissible dose limit of 0.05 Sv (5 rems). The distribution of non-radioactive diiodine tablets to the affected population to saturate the thyroid with iodine was decided at a late stage.

References

[ABE 08] ABELL T.L., CAMILLERI M., DONOHOE K. et al., "Consensus recommendations for gastric emptying scintigraphy: A joint report of the American Neurogastroenterology and Motility Society and the Society of Nuclear Medicine", *Journal of Nuclear Medicine Technology*, vol. 36, pp. 44–54, 2008.

[AGE 09] AGENCE NATIONALE DE SÉCURITÉ DU MÉDICAMENT ET DES PRODUITS DE SANTÉ, FDG Scan (^{18}F) 260 MBq/mL, solution injectable, available at: http://agence-prd.ansm.sante.fr/php/ecodex/rcp/R0157439.htm, 2009.

[AGE 11] AGENCE NATIONALE DE SÉCURITE DU MÉDICAMENT. Résumé des caractéristiques du produit, available at: http://agence-prd.ansm.sante.fr, 2011.

[AGE 13a] AGENCE FRANCE-PRESSE, Une personne âgée sur trois meurt avec Alzheimer aux États-Unis, available at: https://www.lapresse.ca/vivre/sante/201303/19/01-4632601-alzheimer-aux-etats-unis.php, 2013.

[AGE 13b] AGENCE NATIONALE DE SÉCURITÉ DU MÉDICAMENT, Iodure (^{131}I) de sodium pour thérapie, available at: http://agence-prd.ansm.sante.fr/php/ecodex/rcp/R0229270.htm, 2013.

[AGE 14] AGENCE NATIONALE DE SÉCURITÉ DU MÉDICAMENT, EDICIS 2 mg, trousse pour préparation radiopharmaceutique, available at: http://agence-prd.ansm.sante.fr/php/ecodex/rcp/R0235010.htm, 2014.

[AGE 15] AGENCE NATIONALE DE SÉCURITÉ DU MÉDICAMENT, TECHNESCAN SESTAMIBI 1 mg, trousse pour préparation radiopharmaceutique, available at: http://agence-prd.ansm.sante.fr/php/ecodex/rcp/R0254027.htm, 2015.

[AGE 17] AGENCE NATIONALE DE SÉCURITÉ DU MÉDICAMENT, TECHNESCAN DMSA, poudre pour solution injectable. Trousse pour la préparation de la solution injectable de succimère de technétium [99mTc], available at: https://base-donnees-publique.medicaments.gouv.fr/, 2017.

[AGE 19] AGENCE INTERNATIONALE DE SÉCURITÉ DU MÉDICAMENT. TECHNESCAN DTPA, trousse pour la préparation de la solution injectable de pentétate de technétium [99m Tc], available at: https://base-donnees-publique.medicaments.gouv.fr/affichageDoc.php?specid=64060864&typedoc=R, 2019.

[AGE 20a] AGENCE NATIONALE DE LA RECHERCHE (ANR), "Les maladies neurodégénératives : le défi des neurosciences", *Les cahiers de l'ANR*, no. 13, available at: https://anr.fr/fileadmin/documents/2021/ANR_Cahiers_N13_complet.pdf, 2020.

[AGE 20b] AGENCE NATIONALE DE SÉCURITÉ DU MÉDICAMENT, CHLORURE DE THALLIUM [201 Tl], available at: https://base-donnees-publique.medicaments.gouv.fr/, 2020.

[AGE 21] AGENCE NATIONALE DE SÉCURITÉ DU MÉDICAMENT, CERESTAB 500 microgrammes, trousse radiopharmaceutique, available at: https://base-donnees-publique.medicaments.gouv.fr/, 2021.

[AGO 11] AGOSSA K., SANTUCCI J., QUERELLOU S. et al., "Intérêt de la scintigraphie au technétium 99 métastable (^{99m}Tc) pour la prise en charge des ostéonécroses liées aux bisphosphonates", *Médecine buccale chirurgie buccale*, vol. 17, no. 3, pp. 203–209, doi: 10.1051/mbcb/2011115, 2011.

[ALA 08] ALAYA M.B., Synthèse et étude de la biodistribution d'une nouvelle molécule marquée par le technétium-99m, Final studies project, National Institute of Applied Science and Technology, Tunis, 2008.

[ALA 16] AL-ALAWY I.T., MOUHAMED R.S., "Nuclear data relevant to the production of radioiodine I-123, I-125 by indirect route and medical applications", *International Letters of Chemistry, Physics and Astronomy*, vol. 63, pp. 90–97, doi: 10.18052/www.scipress.com/ILCPA.63.90, 2016.

[ALD 48] ALDRICH L.T., NIER A.O., "Argon 40 in potassium minerals", *Physical Review*, vol. 74, pp. 876–877, 1948.

[ALZ 15] ALZHEIMER'S DISEASE INTERNATIONAL, Résumé du rapport mondial Alzheimer 2015 de l'ADI, available at: https://www.alzint.org/u/world-alzheimer-report-2015-summary-sheet-french.pdf, 2015.

[AND 18] ANDREI T., Evaluation d'un ylure d'iodonium et d'un dérivé organostannique en vue de la synthèse de la [^{18}F] fluorobenzylamine, Master's thesis, Université de Liège, Liège, 2018.

[AND 22] ANDRIOLE G.L., Prostatite, available at: https://www.msdmanuals.com/fr/professional/troubles-g%C3%A9nito-urinaires/maladies-b%C3%A9nignes-de-la-prostate/prostatite, 2022.

[ANK 22] ANKOMAH E., Afrique subsaharienne : la charge du cancer devrait presque doubler dans les 20 prochaines années, available at: https://news.un.org/fr/story/2022/05/1119792, 2022.

[ANS 16] ANSQUER C., KRAEBER-BODERE F., Techniques de médecine nucléaire pour l'imagerie et le traitement des tumeurs neuroendocrines gastro-entéro-pancréatiques, in *Radiologie et imagerie médicale : abdominale – digestive*, Elsevier, Amsterdam, doi: 10.1016/S1879-8527(16)67020-X.inserm-02913623, 2016.

[APP 78] APPLEBY P.G., OLDFIELD F., "The calculation of the lead-210 dates assuming a constant rate of supply of unsupported ^{210}Pb to the sediment", *CATENA*, vol. 5, pp. 1–8, 1978.

[APP 01] APPLEBY P.G., "Chronostratigraphic techniques in recent sediments", in LAST W.M., SMOL J.P. (eds), *Tracking Environmental Change Using Lake Sediments*, Kluwer Academic Publishers, The Netherlands, 2001.

[ARD 06] ARDISSON V., Evaluation de nouveaux radiopharmaceutiques, PhD thesis, Université Joseph Fourier-Grenoble 1, Grenoble, 2006.

[ARG 20] ARGENCE J., Graisse viscérale : 4 choses à savoir sur cette mauvaise graisse abdominale, available at: https://www.femmeactuelle.fr/sante/sante-pratique/graisse-viscerale-4-2099921, 2020.

[ASS 20] ASSOCIATION DE RECHERCHE SUR LES CANCERS, Épidémiologie des cancers, available at: https://www.arcagy.org/infocancer/le-cancer/epidemiologie-du-cancer.html/, 2020.

[ASS 21] ASSOCIATION POUR LA RECHERCHE SUR LE CANCER, Cancers de la thyroïde : les symptômes et le diagnostic, available at: https://www.fondation-arc.org/cancer/cancer-thyroide/symptomes-diagnostic-cancer, 2021.

[AUS 11] AUSTRALIAN NUCLEAR SCIENCE AND TECHNOLOGY ORGANIZATION (ANSTO), Production and decay of radioisotopes, available at: https://www.ansto.gov.au.media, 2011.

[AUT 09] AUTIWA, PHQ954 Nucléosynthèse primordiale stellaire, available at: http://autiwa.free.fr/cours/sem9/PHQ954_Nucleosynthese_primordiale_stellaire.pdf, 2009.

[BAI 08] BAILLY S., Le gravitino, candidat à la matière noire et les implications en nucléosynthèse primordiale, PhD thesis, Université Montpellier II, Montpellier, 2008.

[BAR 56] BARNES J.W., LANG E.J., POTRATZ H.A., "Ratio of ionium to uranium in coral limestone", *Science*, vol. 124, pp. 175–176, 1956.

[BAR 95] BARO J., SEMPAU J., FERNÁNDEZ-VAREA J.M. et al., "PENELOPE: An algorithm for Monte Carlo simulation of the penetration and energy loss of electrons and positrons in matter", *Nuclear Instruments and Methods in Physics Research Section B: Beam Interactions with Materials and Atoms*, vol. 610, pp. 31–46, 1995.

[BAS 17] BASTIANETTO S., La maladie de Parkinson, Institut universitaire de gériatrie de Montréal (IUGM), available at: https://www.passeportsante.net/fr/Maux/Problemes/Fiche.aspx?doc=maladie_parkinson_pm, 2017.

[BEC 10] BECK P., Le système solaire, Lecture, M2 SVT 2010–2011, Laboratoire de Planétologie de Grenoble, Grenoble, available at: https://ipag.osug.fr/~beckp/Research/Teaching_files/Planeto_CAPES_2010.pdf, 2010.

[BEK 06] BEKAERT V., Développement d'un tomographe à emission monophotonique dédié au petit animal, Thesis, Université Louis Pasteur - Strasbourg I, Strasbourg, 2006.

[BÉL 16] BÉLIVEAU R., GRINGAS D., *Les aliments contre le cancer*, Les Éditions du Trécarré, Montreal, available at: https://ia803003.us.archive.org/14/items/lesalimentscontrelecancer/%5B%20, 2016.

[BEL 18] BELOUALI-DIALLO S., Optimalisation radiopharmaceutique de la scintigraphie gastrique pour un meilleur accès au diagnostic de la gastroparésie, PhD thesis, Aix-Marseille Université, Marseille, 2018.

[BÊM 18] BÊME D., Calcitonine (thyrocalcitonine), available at: https://www.doctissimo.fr/html/sante/analyses/sa_365_calcitonine, 2018.

[BÊM 19] BÊME D., Les accidents vasculaires cérébraux en chiffres, available at: https://www.doctissimo.fr/html/dossiers/avc/sa_4720_avc_chiffres.htm, 2019.

[BER 21] BERTHON M., Maladie de Parkinson, available at: https://www.deuxiemeavis.fr/pathologie/maladie-de-parkinson, 2021.

[BLA 17] BLAIRE T., Tomoscintigraphie myocardique double-isotope ($^{123}I/^{99m}Tc$) sur gamma-caméra à semi-conducteur : aspects méthodologiques et applications cliniques, PhD thesis, Université Caen Normandie, Caen, 2017.

[BLA 18] BLANC-BEGUIN F., LE ROUX P.-Y., Scintigraphie pulmonaire pour le diagnostic de l'embolie pulmonaire, Lecture, APHO, Université de Bretagne Occidentale, Brest, available at: https://www.apho.fr/wp-content/uploads/2018/04/Scintigraphie-pulmonaire-pour-le-diagnostique-de-lEP.pdf, 2018.

[BLO 46] BLOCH F., "Nuclear induction", *Physical Review*, vol. 70, pp. 4604–4673, 1946.

[BOI 17] BOISSON T., Qu'est-ce que la limite de Chandrasekhar ?, available at: https://trustmyscience.com/qu-est-ce-que-la-limite-de-chandrasekhar/, 2017.

[BOI 18] BOISSON T., Quel est l'âge de l'univers et comment est-il déterminé ?, available at: https://trustmyscience.com/age-de-l-univers-comment-est-il-determine/, 2018.

[BOM 12] BOMBARDIER P., Compréhension et modélisation des émissions environnementales d'aérosols radioactifs liées à l'utilisation d'un générateur de technétium en service de médecine nucléaire, PhD thesis, Université Paris-Diderot – Paris VII, Paris, 2012.

[BON 10] BONNAUD G., L'aérosolthérapie, Polyclinique de Courlancy – Reims, available at: http://pneumocourlancy.fr/page_aerosoltherapie.html, 2010.

[BOT 17] BOTTS S., ALANIS B., KUAI D., Technetium (Tc-99m) Imaging Agents, Worksheet, Texas A&M University, Texas, available at: https://www.chem.tamu.edu/2017%20Student%20Posters/Technetium.pdf, 2017.

[BOU 17] BOUTON L., Médicaments radiopharmaceutiques émetteurs de positions : réglementation et production, PhD thesis, Université de Lille 2, Lille, 2017.

[BOU 18] BOUR H., Veines et artères : quelles différences ?, available at: https://www.topsante.com/medecine/circulation-sanguine/jambes-lourdes/veines-et-arteres-quelles-differences-252571, 2018.

[BOU 20] BOUR H., L'arthrose et l'arthrite sont deux maladies de la famille des rhumatismes, available at: https://www.topsante.com/medecine/rhumatismes/arthrose/arthrose-et-arthrite-quelle-difference-252719, 2020.

[BOY 22] BOYCE H., HAGGARD D., WITZEL G. et al., "Multiwavelength variability of Sagittarius A* in 2019 July", *The Astrophysical Journal*, vol. 931, no. 1, doi: 10.3847/1538-4357/ac6104, 2022.

[BRE 00] BREISMEISTER J.F., MCNP: A general Monte Carlo N-particle transport code, Manual, Version 4C, LA-13709-M, 2000.

[BRÉ 06] BRÉMOND A., "À l'aube de la découverte de l'expansion de l'Univers", *Histoire & mesure*, vol. XXI–2, pp. 157–186, available at: https://journals.openedition.org/histoiremesure/1752, 2006.

[BRÉ 08] BRÉMOND A., Vesto Melvin Slipher (1875–1969) et la naissance de l'astrophysique extragalactique, PhD thesis, Université Claude Bernard – Lyon I, Lyon, 2008.

[BRI 19] BRITISH BROADCASTING CORPORATION, J'ai tout vendu pour payer mon traitement contre le cancer, available at: https://www.bbc.com/afrique/region-50545786, 2019.

[BRO 22] BRON J., Adénome de la prostate, available at: https://docteur-bron-urologue.fr/adenome-de-la-prostate/, 2022.

[BUI 19] BUITEKANT E., Cancer de la thyroïde : symptômes, stades, marqueurs, available at: https://sante.journaldesfemmes.fr/fiches-maladies/2538290-cancer-de-la-thyroide, 2019.

[BUI 21] BUITEKANT E., Quelle est la différence entre une bactérie et un virus ?, available at: https://www.geo.fr/environnement/quelle-est-la-difference-entre-une-bacterie-et-un-virus-204526, 2021.

[BUR 57] BURBIDGE E.M., BURBIDGE G.R., FOWLER W.A., "Synthesis of the elements in stars", *Reviews of Modern Physics*, vol. 29, p. 547, doi: 10.1103/RevModPhys.29.547, 1957.

[BUS 13] BUSKULIC D., Notes de cours de PHYS 801. Introduction à la Physique Nucléaire, https://lappweb.in2p3.fr/~buskulic/cours/PHYS801/PHYS801_Physique_Nucleaire.pdf, 2013.

[BUV 09] BUVAT I., Imagerie en médecine nucléaire, Imagerie et modélisation en neurobiologie et cancérologie, Lecture, Université Paris 11, Orsay, 2009.

[CAR 17] CARDENAS J., La scintigraphie ou scintiscan pulmonaire, available at: https://www.doctissimo.fr/sante/imagerie-medicale/scintigraphie-pulmonaire, 2017.

[CEN 19] CENTRE HOSPITALIER UNIVERSITAIRE VAUDOIS, Anatomie et physiologie, available at: https://www.chuv.ch/fr/transplantation/anatomie-et-physiologie, 2019.

[CEN 20] CENTRE HOSPITALIER UNIVERSITAIRE VAUDOIS, Médecine nucléaire, available at: https://www.chuv.ch/fr/medecine-nucleaire, 2020.

[CEN 21a] CENTRE D'EXPLORATIONS ISOTOPIQUES, Passer un PET scan, qu'est-ce que cela signifie ? Pourquoi passer un TEP scan ?, available at: https://www.scintigraphie-tep.fr/qu-est-ce-qu-un-tep-scan-ou-petscan.php, 2021.

[CEN 21b] CENTRE D'EXPLORATIONS ISOTOPIQUES, Passer une scintigraphie rénale (DMSA ou DTPA/MAG3) à Rennes ou Saint Malo, available at: https://www.scintigraphie-tep.fr/le-deroulement-de-la-scintigraphie-renale.php, 2021.

[CEN 21c] CENTRE DE RADIOLOGIE ET D'IMAGERIE MÉDICALE BARTHOLDI, Radiologie avec produit de contraste, available at: https://www.radiologiebartholdi.com/radiologie-produit-contraste.html, 2021.

[CEN 22] CENTRE DE CANCEROLOGIE LES DENTELLIÉRES, La chimiothérapie dans le traitement du cancer de la prostate, available at: https://www.elsan.care/fr/centre-cancerologie-dentellieres/nos-actualites/la-chimiotherapie-dans-le-traitement-du-cancer-de, 2022.

[CEY 19] CEYRAT Q., BORDENAVE L., COUFFINHAL T. et al., "Évaluation quantitative de la réserve coronaire avec les caméras CZT, faisabilité et perspectives", *Médecine Nucléaire*, vol. 43, no. 2, p. 185, 2019.

[CHA 10] CHARLASSIER R., Mesure des anisotropies de polarisation du fond diffus cosmologique avec l'interféromètre bolométrique QUBIC, PhD thesis, Université Paris Diderot-Paris 7, Paris, 2010.

[CHA 16] CHARAF C.M., Cours 2. Formation des éléments chimiques, Université Ferhat Abbas, Sétif1, Institut d'Architecture et des Sciences de la Terre, 2015–2016.

[CHA 19] CHAINE C., Accident ischémique transitoire, available at: https://www.docteurclic.com/maladie/accident-ischemique-transitoire.aspx, 2019.

[CHA 21] CHARLINE D., Ostéomyélite, Santé sur le net, available at: https://www.sante-sur-le-net.com/maladies/maladies-infectieuses/osteomyelite/, 2021.

[CHA 22] CHANDRASEKAR T., Cancer de la prostate, available at: https://www.msdmanuals.com/fr/professional/cancer-de-la-prostate, 2022.

[CHE 55] CHERDYNTSEV V.V., CHALOV P.I., KHITRIK M.E., On isotopic composition of radioelements in naturel objects and problems of geochronology, in *Determ. Abs. Age Geol. Formns. Izd. Akad. Nauk* (in Russian), SSSR, Moscow, 1955.

[CHO 09] CHOPRA A., Molecular imaging and contrast agent database (MICAD), National Center for Biotechnology Information, NLM, NIH, Bethesda, MD, 2009.

[CHO 20a] CHOI S.K., HASSELFIELD M., PATTY HO S.-P. et al., "The Atacama cosmology telescope: A measurement of the Cosmic Microwave Background power spectra at 98 and 150 GHz", *JCAP12045*, doi: 10.1088/1475-7516/2020/12/045, 2020.

[CHO 20b] CHONG J.Y., Présentation de l'accident vasculaire cérébral, available at: https://www.msdmanuals.com/fr/accueil/troubles-du-cerveau,-/accident-vasculaire-, 2020.

[CIM 79] CIMETIERE C., LEGER J., MUXART R. et al., Utilisation de cyclotrons pour la préparation de radionucléides à usage médical, in *Seminar on Particle Accelerators: Use – Radioprotection*, CBA – CONF 4641, Saclay, 14–19 May 1979.

[CIM 17] CIMARELLI S., MINGECO O., PERRIN-FAYOLLE O. La scintigraphie des parathyroïdes. Médecine nucléaire de la Doua, available at: https://www.medecine-nucleaire.fr/scintigraphie-parathyroides.html, 2017.

[COH 77] COHEN-TANOUNDJI C., DIU B., LALOE F., *Mécanique quantique 1*, Hermann Éditeurs des Sciences et des Arts, Paris, 1977.

[COH 92] COHEN-TANOUNDJI C., DIU B., LALOE F., *Mécanique quantique 2*, Hermann Éditeurs des Sciences et des Arts, Paris, 1992.

[COL 16] COLLEGE UNIVERSITAIRE DES ENSEIGNANTS EN NEPHROLOGIE, Filtration glomérulaire et débit sanguin rénal, available at: http://www.cuen.fr/lmd/spip.php?rubrique94, 2016.

[COL 19] COLEIRO A., L'univers de haute énergie, Lecture, Laboratoire APC/Université Paris Diderot, Paris, available at: http://html5.ens-lyon.fr/Acces/Astronomie/20190820/20190820_LUniversDeHauteEnergie_1.pdf, 2019.

[COM 12] COMMISSARIAT À L'ÉNERGIE ATOMIQUE, Mesurer l'âge de l'univers, available at: http://bonnetbidaud.free.fr/pedagogie/hubble_law/web_database/Age_de_univers.pdf, 2012.

[COM 13] COMMENT A., Résonance magnétique nucléaire : théorie et manuel pratique, Practical study, École Polytechnique Fédérale de Lausanne, Lausanne, available at: https://docplayer.fr/15610697-Resonance-magnetique-nucleaire-theorie-et-manuel-pratique.html, 2013–2014.

[COM 15] COMMISSARIAT À L'ÉNERGIE ATOMIQUE, Nucléosynthèse primordiale, available at: https://www.cea.fr/comprendre/Pages/matiere-univers/astrophysique-nucleaire.aspx?Type=Chapitre&numero=3, 2015.

[COM 16] COMMISSION CANADIENNE DE SÛRÉTE NUCLÉAIRE, Imagerie médicale et radiothérapie, available at: https://www.cnsc-ccsn.gc.ca/fra/resources/infographics/mir/index.cfm, 2016.

[COM 17] COMMISSARIAT À L'ÉNERGIE ATOMIQUE, Les étoiles, available at: https://www.cea.fr/comprendre/Pages/matiere-univers/essentiel-sur-les-etoiles.aspx, 2017.

[COM 18] COMMISSARIAT À L'ÉNERGIE ATOMIQUE ET AUX ÉNERGIES ALTERNATIVES, Les applications de la radioactivité, available at: https://www.cea.fr/comprendre/Pages/radioactivite.aspx?Type=Chapitre&numero=3, 2018.

[COM 22] COMMISSION CANADIENNE DE SÛRÉTE NUCLÉAIRE, Effets de l'accident de Tchernobyl sur la santé, available at: https://nuclearsafety.gc.ca/fra/resources/health/health-effects-chernobyl-accident.cfm, 2022.

[CON 21] CONSORTIUM NATIONAL DE FORMATION EN SANTÉ – VOLET UNIVERSITÉ D'OTTAWA, Accident vasculaire cérébral, available at: https://cnfs.ca/pathologies/accident-vasculaire-cerebral-ischemique-et-hemorragique, 2021.

[COR 08] CORVOL P., "VEGF, anti-VEGF et pathologies", *Bulletin de l'Académie nationale de médecine*, vol. 192, no. 2, pp. 289–302, 2008.

[COU 18] COULON J.-P., Production et utilisation des radioéléments, in *6ème séminaire : médecine et nucléaire*, CEA Cadarache, available at: https://docplayer.fr/79889717-Production-et-utilisation-des-radioelements.html, 2018.

[DAI 13] DAIGLE S.M., Low energy proton capture study of the $^{14}N(p, \gamma)^{15}O$ reaction, PhD thesis, University of North Carolina at Chapel Hill, Chapel Hill, doi: 10.17615/nhbv-w434, 2013.

[DAR 21] DARMON I., La chimiothérapie dans le traitement du cancer, available at: https://radiotherapie-hartmann.fr/actualites/traitements/chimiotherapie/, 2021.

[DEB 08] DE BECO V., LE BARS D., SCHERRMANN J.-.M., "Le fluor 18 en radiopharmacie. Fluorine-18 in radiopharmacy", *Annales pharmaceutiques françaises*, vol. 66, no. 1, pp. 60–65, 2008.

[DEC 19] DECLERCK P., Scintigraphie cérébrale (HMPAO-ECD), available at: https://www.clstjean.be/sites/default/files/brochures/services/medecine_nucleaire/, 2019.

[DED 04] DE DREUILLE O., MASZELIN P., FOEHRENBACH H. et al., "Principe et technique de la tomographie par émission de positons (TEP)", *EMC-Radiologie*, vol. 1, pp. 2–35, doi: 10.1016/S1762-4185(03)00006-2, 2004.

[DEG 17] DEGRELLE D., Caractérisation numérique de la technique de spectrométrie gamma par simulation Monte-Carlo. Application à la datation d'échantillons environnementaux, PhD thesis, Université Bourgogne Franche Comté, Besançon, 2017.

[DEL 17] DELBECQ D., La foudre déclenche des réactions… nucléaires, available at: https://www.lemonde.fr/sciences/la-foudre-declenche-des-reactions-nucleaires, 2017.

[DEL 19] DELLA VALLE A.-C., Scintigraphie : définition et déroulé de l'examen, available at: https://sante.journaldesfemmes.fr/fiches-anatomie-et-examens/, 2019.

[DEL 23] DE LA VAISSIERE C., LABERRIGUE-FROLOW J., SACQUIN Y. et al., Retombées essais nucléaires, available at: https://laradioactivite.com/energie_nucleaire/retombees_essais_nucleaires, 2023.

[DÉP 22] DÉPARTEMENT PREVENTION CANCER ENVIRONNEMENT, Sarcome, available at: https://www.cancer-environnement.fr/fiches/cancers/sarcome/, 2022.

[DES 00] DESUZINGES D., *Les radiopharmaceutiques et la radiopharmacie : aspects réglementaires et techniques*, École Normale de la Santé Publique, Rennes, 2000.

[DES 13] DESCOTEAUX C., Développement de nouveaux composés anticancéreux pour le traitement des cancers féminins, PhD thesis, Université du Québec à Trois Rivières (UQTR), Quebec, 2013.

[DES 15] DESCOURS N., HECQUET N., LEPART C., Étude rétrospectives de la prise en charge pluridisciplinaire de la maladie de Basedow sévère au sein du service de neuro-ophtalmologie de l'hôpital Pierre Wertheimer, Orthoptist qualification certificate, Université Claude Bernard Lyon 1, 2015.

[DEV 53] DE VRIES H., BARENDSEN G.W., "Radiocarbon dating by a proportional counter filled with carbon dioxide", *Physica (Amsterdam)*, vol. 19, p. 987, 1953.

[DIE 22] DIENG M.D., Cancer de la prostate, radioscopie d'une tumeur fatale aux hommes, available at: https://www.seneplus.com/sante/radioscopie-dune-tumeur-fatale-aux-hommes, 2022.

[DIL 92] DILLENSEGER J.-L., Imagerie tridimensionnelle morphologique et fonctionnelle en multimodalité. Traitement du signal et de l'image, PhD thesis, Université de Tours, Tours, 1992.

[DIO 74] DIOP C.A., *Physique Nucléaire et chronologie absolue*, IFAN/Nouvelles Éditions Africaines, Dakar, 1974.

[DIR 99] DIRECTION GÉNÉRALE ENVIRONNEMENT, Radioprotection 97. Radioprotection à la suite d'une thérapie à l'iode 131, available at: https://op.europa.eu/fr/publication-detail/-/publication/, 1999.

[DIS 14] DI SALVIO A., Méthode efficace d'assignation de tissus humains par tomodensitométrie à double énergie, Master's thesis, Université de Montréal, Montreal, 2014.

[DOL 06] DOLLÉ F., PERRIO C., BARRÉ L. et al., "Les molécules marquées au fluor-18. Synthèse et application en imagerie médicale", *L'actualité chimique*, nos 301–302, 2006.

[DOM 18] DOMINGUEZ E., La démence à corps de Lewy, PhD thesis, Université de Picardie Jules Verne, Amiens, 2018.

[DON 76] DONALD W., GAULT C.H., "Prediction of creatinine clearance from serum creatinine", *Nephron*, vol. 16, no. 1, pp. 31–41, doi: 10.1159/000180580, 1976.

[DOU 20] DOUTRELEAU S., CHEVALIER L., TABET J.-Y. et al., "Mise au point sur un examen essentiel", *Revue Cardio & Sport*, no. 45, 2020.

[DUC 15] DUCHEMIN C., Étude de voies alternatives pour la production de radionucléides innovants pour les applications médicales, PhD thesis, Université de Nantes, Nantes, 2015.

[DUP 01] DUPONT Y., Distance, mouvement, masse et rayon des étoiles, available at: https://craq-astro.ca/phy1971/chap_pdf/chap15b.pdf, 2001.

[DÜR 15] DÜRRLEMAN J., Valeur diagnostique de la mammographie synthétique sans et avec tomosynthèse, PhD thesis, Université Paris Descartes, Paris, 2015.

[ÉCO 13] ÉCOLE DE TECHNOLOGIE SUPÉRIEURE, GTS503_C4-Médecine Nucléaire, available at: https://cours.etsmtl.ca/gts503/Cours/GTS503_C4-ImagerieNucleaire.pdf, 2013.

[ESC 22] ESCULAPE. Les prostatites : types, causes, symptômes, complications, diagnostic, traitements, available at: https://www.esculape.com/uronephro/prostatite.html, 2022.

[ESS 19] ESSLINGER O., La nucléosynthèse primordiale, available at: https://www.astronomes.com/le-big-bang/nucleosynthese-primordiale, 2019.

[EUS 15] EUSTACHE P., Evaluation par IRM multimodale des modifications cérébrales chez des patients Alzheimer à un stade prodromique. Optimisation de la relaxométrie T2* par IRM, PhD thesis, Université de Toulouse, Toulouse, 2015.

[FAR 17] FARRELL J., The tangled history of big bang science, available at: https://nautil.us/blog/the-tangled-history-of-big-bang-science, 2017.

[FÉD 20] FÉDÉRATION FRANÇAISE DE CARDIOLOGIE, L'infarctus du myocarde, available at: https://www.fedecardio.org/Les-maladies-cardio-vasculaires/linfarctus-du-myocarde, 2020.

[FER 04] FERNANDO M.S., INCE P.G., "Vascular pathologies and cognition in a population based cohort of elderly people", *Journal of the Neurological Sciences*, vol. 226, pp. 13–17, doi: 10.1016/j.jns.2004.09.004, 2004.

[FER 07] FERRER M., SUÁREZ J.F., GUEDEA F. et al., "Health-related quality of life 2 years after treatment with radical prostatectomy, prostate brachytherapy, or external beam radiotherapy in patients with clinically localized prostate cancer", *International Journal of Radiation Oncology Biology Physics*, vol. 72, no. 2, pp. 421–432, 2007.

[FER 11] FERRER L., Dosimétrie clinique en radiothérapie moléculaire : optimisation de protocoles et implémentation clinique, PhD thesis, Université de Nantes, Nantes, 2011.

[FER 19] FERRARI J., La liste des anti-inflammatoires stéroïdiens, available at: https://sante.journaldesfemmes.fr/fiches-sante-du-quotidien/2542984-la-liste-des-anti-inflammatoires-steroidiens/, 2019.

[FEY 19] FEYROUZ L., MEBARKA L., Étude de la réalisation d'un détecteur de rayons gamma, Master's thesis, Université Abou Bekr Belkaïd de Tlemcen, Tlemcen, 2019.

[FON 04] FONTUGNE M., "Les derniers progrès du calibrage des âges radiocarbone permettent-ils une révision des chronologies entre 25 et 50.000 ans B.P. ?", *Quaternaire*, vol. 15, no. 3, pp. 245–252, doi: 10.3406/quate.2004.1771, 2004.

[FON 20] FONDATION CONTRE LE CANCER, Tomographie par émission de positrons (TEP), available at: https://www.cancer.be/les-cancers/le-pet-scan-imagerie-m-tabolique, 2020.

[FOX 71] FOX S.M., NAUGHTON J.P., HASKELL W.L., "Physical activity and the prevention of coronary heart disease", *Annals of Clinical Research*, vol. 3, no. 6, pp. 404–432, 1971.

[FRA 07] FRANK S., PISTERS L.L., DAVIS J. et al., "An assessment of quality of life following radical prostatectomy, high dose external beam radiation therapy and brachytherapy iodine implantation as monotherapies for localized prostate cancer", *The Journal of Urology*, vol. 177, no. 6, pp. 2151–2156, 2007.

[FRA 09] FRAJNDLICH R., "IEA-R1 research reactor: Operational life extension and considerations regarding future decommissioning", in *International Nuclear Atlantic Conference – INAC 2009*, Rio de Janeiro, 2009.

[FRA 14] FRANÇOIS A., Élaboration de complexes hétérobinucléaires par approche chimie. Click pour une application en imagerie bimodale, PhD thesis, Université de Toulouse, Toulouse, 2014.

[FRE 01] FREEDMAN W.L., MADORE B.F., GIBSON B.K. et al., "Final results from the hubble space telescope key project to measure the hubble constant", *Astrophysical Journal*, vol. 553, pp. 47–72, doi: 10.1086/320638, 2001.

[FRE 14] FREER M., FAYNBO H., "The Hoyle state in 12C", *Progress in Particle and Nuclear Physics*, vol. 78, pp. 1–23, doi: 10.1016/j.ppnp.2014.06.001, 2014.

[FRU 18] FRUET G., Structure des ions lourds et nucléosynthèse dans les étoiles massives : la réaction $^{12}C + ^{12}C$, PhD thesis, Université de Strasbourg, Strasbourg, 2018.

[FUS 72] FUSCO M.A., PEEK N.F., JUNGERMAN J.A. et al., "Cyclotron production of carrier-free 123I using the reaction 127I (p, 5n) 123Xe", *The Journal of Nuclear Medicine*, vol. 13, no. 10, pp. 729–732, 1972.

[GAL 04] GALALAE R., MARTINEZ A., MATE T. et al., "Long-term outcome by risk factors using conformal high-dose-rate brachytherapy (HDR-BT) boost with or without neoadjuvant androgen suppression for localized prostate cancer", *International Journal of Radiation Oncology*, vol. 58, no. 4, pp. 1048–1055, 2004.

[GAR 75] GARNER E.L., MURPHY T.J., GRAMLICH J.W. et al., "Absolute isotopic abundance ratios and the atomic weight of a reference sample of potassium", *Journal of Research of the National Bureau of Standards*, vol. 79A, no. 6, pp. 713–725, 1975.

[GAS 18] GASAGHRANDE K., 68Ga-PSMA-11, nouveau traceur TEP pour l'imagerie du cancer de la prostate : synthèse, contrôles qualité et dossier d'autorisation, PhD thesis. Université Toulouse III, Toulouse, 2018.

[GIO 19] GIORGETTA J., PET scan : pourquoi en faire, quels effets secondaires après ?, available at: https://sante.journaldesfemmes.fr/fiches-anatomie-et-examens/, 2019.

[GIO 20] GIORGETTA J., Ganglion lympathique : symptômes, causes et traitement, available at: https://sante.journaldesfemmes.fr/2630995-ganglion-lympathique-cou-aisselle-/, 2020.

[GIO 21] GIORGETTA J., Capillaire sanguin : définition, rôle, schéma, fragilité, available at: https://sante.journaldesfemmes.fr/fiches-anatomie-et-examens/, 2021.

[GOU 10] GOULART D.M., Imagerie médicale, available at: https://scinti.edu.umontpellier.fr/files/2010/08/INITIATION-A-LIMAGERIE-MEDICALE.pdf, 2010.

[GRO 15] GROUPEMENT D'INTÉRÊT PUBLIC (GIP CYROI), Production de radiopharmaceutiques. Développement de la recherche en radiochimie et en imagerie, available at: https://www.asn.fr, 2015.

[GRO 19] GROUSSIN L., Quelle vie après chirurgie cervicale (thyroïdienne et parathyroïdienne), available at: https://www.reseau-gte.org/2019/12/3_vie-apres-chirurgie-tne_3.pdf, 2019.

[GUÉ 19] GUÉRIN M., Infiltration, crème à la cortisone : les précautions, available at: https://sante.journaldesfemmes.fr/cortisone-infiltration-creme-precautions/, 2019.

[GUI 17] GUIBBAL F., Développement, synthèse et marquage au fluor-18 de ligands de la Lp-PLA2 pour la détection précoce de plaques d'athérome par imagerie TEP, PhD thesis, Université de La Réunion, La Réunion, 2017.

[HAB 13] HABERT M.O., KAS A., "Aspects techniques et pratiques, et indications de la tomographie d'émission monophotonique cérébrale", *EMC-Neurologie*, vol. 10, no. 3, doi: 10.1016/S0246-0378(13)52780-1, 2013.

[HAM 90] HAMON M., PELLERIN F., GUERNET M. et al., *Chimie analytique, tome 3 : méthodes spectrales et analyse organique*, Elsevier Masson, Paris, 1990.

[HAM 14] HAMAGUCHI T., MYIATI T., MATSUSHITA T. et al. "Analysis of spatial dependance of acoustic noisetransfer function in magnetic resonance imaging", *European Congress of Radiology*, C-1988, doi: 10.1594/ecr2014/C-1988, 2014.

[HAN 16] HANANE A., Conception et réalisation d'une banque de données médicale, Master's thesis, Université Abou Bakr Belkaïd de Tlemcen, Tlemcen, 2016.

[HAU 19] HAUTE AUTORITÉ DE SANTÉ. CERESTAB – CERETEC (exametazime), available at: https://www.has-sante.fr/jcms/p_3106649/fr/cerestab-ceretec-exametazime, 2019.

[HAU 22] HAUTE AUTORITÉ DE SANTÉ, Accident vasculaire cérébral. Prescription d'activité physique, available at: https://www.has-sante.fr/upload/docs/application/pdf/2022-08/synthese_aps_avc_vf.pdf, 2022.

[HEC 22] HECTOR A., Tremblement, available at: https://www.msdmanuals.com/tremblement, 2022.

[HEN 19] HENRY A., Qu'est-ce qu'une scintigraphie osseuse ?, available at: https://www.femmeactuelle.fr/sante/sante-pratique/scintigraphie-osseuse, 2019.

[HER 14] HERMESSE J., NICKERS P., REUL O. et al., "La curiethérapie par implants permanents d'iode 125 : résultats après 10 ans d'expérience au CHU de Liège", *Revue médicale de Liège*, vol. 69, supp. I, pp. 32–36, 2014.

[HER 18] HERNADEZ D., Étude de l'influence de l'introduction d'une aiguille d'entrée d'air sur la qualité d'un médicament radiopharmaceutique : le Myoview®, PhD thesis, Université Claude Bernard – Lyon 1, Lyon, 2018.

[HER 19] HERBERT A., Mise au point de nouvelles techniques de radio-iodation et application au radiomarquage de molécules d'intérêt, PhD thesis, Université Caen Normandie, Caen, 2019.

[HÔP 21] HÔPITAUX UNIVERSITAIRES GENÈVE, Qu'est qu'un AVC et un AIT ?, available at: https://www.hug.ch/accident-vasculaire-cerebral/qu-est-qu-avc-ait, 2021.

[HÔP 23a] HÔPITAL PARIS SAINT-JOSEPH, Hémorragie méningée, https://www.hpsj.fr/specialites/neuro-vasculaire/pathologies/hemorragie-meningee/, 2023.

[HÔP 23b] HÔPITAUX UNIVERSITAIRES GENÈVE, Accident vasculaire cérébral. Le diagnostic et le traitement, available at: https://www.hug.ch/accident-vasculaire-cerebral/diagnostic-traitement, 2023.

[HOS 23] HOSPICES CIVILS DE LYON, Hypertrophie bénigne de prostate (Adénome de prostate), available at: https://www.chu-lyon.fr/hypertrophie-benigne-de-prostate-adenome-de-prostate, 2023.

[HUB 29] HUBBLE E., "A relation between distance and radial velocity among extra-galactic nebulae", *Proceedings of the National Academy of Sciences*, vol. 15, no. 3, pp. 168–173, 1929.

[HUB 04] HUBERT P., HUBERT F., RAFFESTIN-TORT V., “La datation des vins : une application des mesures des très faibles radioactivités”, *Bulletin de l'union des physiciens*, vol. 98, no. 862, pp. 381–395, 2004.

[HUR 22] HURST R., COOPER C., CLARK J., Le cancer de la prostate pourrait être lié à des bactéries et cela ouvre la voie à de nouveaux traitements, available at: https://www.ouest-france.fr/le-cancer-de-la-prostate-, 2022.

[IMA 16] IMAIOS, L'IRM pas à pas : cours interactif sur l'imagerie par résonance magnétique, available at: http://www.imaios.com/fr/e-Cours/e-MRI, 2016.

[IMB 05] IMBRIANI G., COSTANTINI H., FORMICOLA A. et al., “S-factor of ^{14}N (p, γ)^{15}O at astrophysical energies”, *European Physical Journal A*, vol. 25, pp. 455–466, doi: 10.1140/epja/i2005-10138-7, 2005.

[INS 01] INSTITUTO DE PESQUISAS ENERGÉTICAS E NUCLEARES (IPEN), Technical Meeting of project counterparts on Cyclotron production of I-123, Sao Paulo, 2001.

[INS 09a] INSTITUT DE RADIOPROTECTION ET DE SÛRETÉ NUCLÉAIRE, Radioprotection : radionucléides. Iode-131, ED 4300, 3rd edition, IRSN, Paris, 2009.

[INS 09b] INSTITUT DE RADIOPROTECTION ET DE SÛRETÉ NUCLÉAIRE, Radioprotection : radionucléides. Iode-123, ED 4303, 2nd edition, IRSN, Paris, 2009.

[INS 10] INSTITUT NATIONAL DU CANCER, Les traitements du cancer de la prostate, available at: https://www.e-cancer.fr, 2010.

[INS 13a] INSTITUT DE RADIOPROTECTION ET DE SÛRETÉ NUCLÉAIRE, Radioprotection : radionucléides. Fluor-18, ED 4311, 1st edition, IRSN, Paris, 2013.

[INS 13b] INSTITUT DE RADIOPROTECTION ET DE SÛRETÉ NUCLÉAIRE, Radioprotection : radionucléides. Thallium-201, ED 4312, 1st edition, IRSN, Paris, 2013.

[INS 18] INSTITUT DE RADIOPROTECTION ET DE SÛRETÉ NUCLÉAIRE, Champs électromagnétiques. L'imagerie par résonance magnétique, ED 4209, 2nd edition, IRSN, Paris, 2018.

[INS 20] INSTITUT DE FORMATION EN SOINS INFIRMIERS (IFSI) DE SAVERNE, La filtration glomérulaire, available at: https://www.ifsi-ifas-saverne.fr/2020/02.pdf, 2020.

[INS 21a] INSTITUT NATIONAL DU CANCER, Anatomie du rein, available at: https://www.e-cancer.fr/Patients-et-proches/Les-cancers/Cancer-du-rein/Anatomie-du-rein, 2021.

[INS 21b] INSTITUT NATIONAL DU CANCER, Données globales d'épidémiologie des cancers, available at: https://www.e-cancer.fr/Professionnels-de-sante/Epidemiologie-des-cancers/Donnees-globales, 2021.

[INS 22] INSTITUT NATIONAL DU CANCER, Quel déroulement ? Curiethérapie, available at: https://www.e-cancer.fr/Cancer-de-la-prostate/Curietherapie/Quel-deroulement, 2022.

[INT 21] INTERNATIONAL ATOMIC ENERGY AGENCY, Virtual Cancer Control Assessment in Senegal by IAEA and Partners, https://www.iaea.org/newscenter/news/virtual-cancer-control-assessment-in-senegal-by-iaea-and-partners, 2021.

[ISN 07] ISNARDI V., KELLY A., BOISSON N. et al., “Comparaison entre les modes d’acquisition 2D et 3D en TEP au FGD”, *Médecine Nucléaire*, vol. 31, no. 10, pp. 538–544, doi: 10.1016/j.mednuc.2007.02.006, 2007.

[JAN 15] JANEVIK E., “Radiopharmaceuticals in neurological and psychiatric disorders”, in *International Conference on Clinical PET-CT and Molecular Imaging (IPET 2015): PET-CT in the Era of Multimodality Imaging and Image-guided Therapy*, Vienna, 5–9 October 2015.

[JIM 19] JIMENEZ C., ERWIN W., CHASEN B., “Targeted radionuclide therapy for patients with metastatic pheochromocytoma and paraganglioma: From low-specific-activity to high-specific-activity iodine-131 metaiodobenzylguanidine”, *Cancers*, vol. 11, p. 1018, doi: 10.3390/cancers11071018, 2019.

[JOY 13] JOYARD Y., Synthèse de nouveaux radiomarqueurs potentiels de l’hypoxie tumorale : développement d’une nouvelle méthodologie de fluoration nucléophile et son application vers la synthèse du 2-[18F]Fluoro-2-désoxy-D-glucose, PhD thesis, INSA de Rouen, Rouen, 2013.

[JUL 03] JULL A.J.T., “Radiocarbon”, in *18th Conference*, vol. 46, Wellington, available at: http://www.radiocarbon.org/, 2003.

[KAH 77] KAHN M., KLEINBERG J., *Radiochemistry of Iodine*, Technical Information Center, Energy Research and Development Administration, Oak Ridge, 1977.

[KAS 11] KASTLER B., VETTER D., “Le phénomène de résonance magnétique”, in *Comprendre l’IRM – Manuel d’auto-apprentissage*, 7th edition, Elsevier Masson, Issy-les-Moulineaux, p. 389, 2011.

[KIR 13] KIRSEBOM O.S., “^{12}C and the triple-α reaction rate”, *Journal of Physics: Conference Series*, vol. 436, p. 012072, doi: 10.1088/1742-6596/436/1/012072, 2013.

[KNI 23] KNIPE H., YAP J., GAILLARD F., Luxury perfusion, doi: 10.53347/rID-57188, 2023.

[LAC 11] LACOEUILLE F., Conception, développement et évaluation d’un nouveau radiopharmaceutique pour la scintigraphie de la perfusion pulmonaire, PhD thesis, Université d’Angers, Angers, 2011.

[LAL 79] LALOU C., HOANG C.-T., “Les méthodes de datation par les descendants de l’Uranium”, *Bulletin de l’Association française pour l’étude du quaternaire*, vol. 16, nos 1–2, pp. 3–14, doi: 10.3406/quate.1979.1341, 1979.

[LAM 72] LAMBREHT R.M., WOLF A.P., “The ^{122}Te (^{4}He, 3n) $^{123}Xe \rightarrow {}^{123}I$”, *Radiation Research*, vol. 52, pp. 32–46, 1972.

[LAM 20] LAMOLINERIE M., Maladie du tremblement essential, available at: https://www.deuxiemeavis.fr/pathologie/maladie-du-tremblement-essentiel, 2020.

[LAP 12] LAPORTE B., Analyse de concordance entre IRM de perfusion de stress et la scintigraphie myocardique à partir des 3 premières années d’expérience nantaise, PhD thesis, Université de Nantes, Nantes, 2012.

[LAS 21] LA SOCIÉTÉ ALZHEIMER DE QUÉBEC, La maladie d'Alzheimer en chiffres, available at: https://www.societealzheimerdequebec.com/comprendre-la-maladie/statistiques/2017, 2021.

[LE 20] LE J., Administration des medicaments, *Le manuel MSD*, available at: https://www.msdmanuals.com/fr/accueil/m%C3%A9dicaments/, 2020.

[LEB 23] LEBON V., Les fondamentaux – chapitre 5 – médecine nucléaire, available at: https://www.cnp-mn.fr/les-fondamentaux-chapitre-5-medecine-nucleaire/, 2023.

[LEH 08] LEHAUTE F., Biochimie des contaminants organiques HAP, PCB et pesticides organochlores dans les sédiments de l'étang de Thau, PhD thesis, Université Pierre et Marie Curie, Paris, 255 p, 2008.

[LEM 27] LEMAITRE G., "Un Univers homogène de masse constante et de rayon croissant rendant compte de la vitesse radiale des nébuleuses extragalactiques", *Annales de la société scientifique de Bruxelles*, vol. A47, pp. 49–59, 1927.

[LEM 23] LE MOAL D., Iode, essentiel pour la thyroïde et le cerveau, available at: https://www.mesbienfaits.com/iode/, 2023.

[LEP 18] LEPRETRE P., Prédiction interictale de la latéralisation du foyer épileptique par imagerie IRM multimodale dans les épilepsies temporales de l'adulte, PhD thesis, Université de Montpellier, Montpellier, 2018.

[LIB 49] LIBBY W.F., ANDERSON E.C., ARNOLD J.R. et al., "Age determination by radiocarbon content: Worldwide assay of natural radiocarbons", *Science*, vol. 109, pp. 227–228, 1949.

[LOM 20] LOMBRISER L., Le mystère de l'expansion de l'Univers trouve une solution, doi: 10.1016/j.physletb.2020.135303, 2020.

[LUM 97] LUMINET J.-P., *L'invention du Big Bang*, Éditions du Seuil, Paris, available at: https://luth.obspm.fr/~luminet/Books/FL.html, 1997.

[MAB 08] MABIT L., BENMANSOUR M., WALLING D.E., "Comparative advantages and limitations of the fallout radionuclides ^{137}Cs, ^{210}Pbex and ^{7}Be for assessing soil erosion and sedimentation", *Journal of Environmental Radioactivity*, vol. 99, pp. 1799–1807, 2008.

[MAG 21] MAGNARD B., SEYNAT M., Réaliser une chromatographie sur couche mince, available at: https://www.maxicours.com/se/cours/realiser-une-chromatographie-sur-couche-mince/, 2021.

[MAL 13] MALLET E., Comment mieux comprendre le métabolisme de la vitamine D ?, available at: https://www.realites-pediatriques.com/, 2013.

[MAL 23] MALVEZZI M., SANTUCCI C., BOFFETTA P. et al., "European cancer mortality predictions for the year 2023 with focus on lung cancer", *Annals of Oncology*, doi: 10.1016/j.annonc.2023.01.010, 2023.

[MAN 20] MANGEANT R., Radiotraceurs des récepteurs sérotoninergiques utilisés chez l'homme, PhD thesis, Université Caen Normandie, Caen 2020.

[MAR 10] MARTIN J.-P., Conférence : le big bang pour les nuls, Cité des Sciences de Paris, available at: https://www.planetastronomy.com/special/2011-special/13nov10/jpm-bb.htm, 2010.

[MAR 12a] MARTA M., The $^{14}N(p,\gamma)^{15}O$ reaction studied at low and high beam energy, available at: https://www.hzdr.de/publications/PublDoc-5915.pdf, 2012.

[MAR 12b] MARTINAGE O., Elaboration d'une nouvelle plateforme pour le développement de traceurs in vivo : application à l'imagerie de la néoangiogenèse tumorale, PhD thesis, Université Paris Sud 11, Paris, 2012.

[MAR 15] MARCADET D.M., ZERBIB E., "Évaluation de l'ischémie Myocardique", *Cardiologie – cardinale*, vol. 9, no. 75, 2015.

[MAR 21] MARTORELL M., Les troubles thyroïdiennes, Institut Wanxiang – 26100 Romans sur Isère, available at: https://www.institut-wanxiang.com/info-sante-mtc.php?edit=59, 2021.

[MED 14] MEDRAN-NAVARRETE V., Synthèse de nouvelles sondes moléculaires marquées au fluor-18 pour l'imagerie de la neuroinflammation par Tomographie par Emission de Positons, PhD thesis, Université Paris-Sud, Paris, 2014.

[MED 18] MEDTRONIC BELGIUM, À propos de la maladie de Parkinson, available at: https://www.medtronic.com/be-fr/patients/pathologies/maladie-de-parkinson.html, 2018.

[MEN 06] MENDOUBI M., La scintigraphie myocardique : technique et indications, PhD thesis, Université Hassan II, Casablanca, 2006.

[MIN 17] MINISTÈRE DES SOLIDARITÉS ET DE LA SANTÉ, STAMICIS 1 mg, trousse pour préparation radiopharmaceutique, available at: https://base-donnees-publique.medicaments.gouv.fr/, 2017.

[MIN 21] MINISTÈRE DES SOLIDARITÉS ET DE LA SANTÉ, PULMOTEC, Creuset de graphite pour la préparation de Technegas pour inhalation, available at: https://base-donnees-publique.medicaments.gouv.fr/, 2021.

[MON 18] MONTEMAGNO C., Développement de radiotraceurs pour l'imagerie phénotypique des cancers du sein métastatiques, Cancer, PhD thesis, Université Grenoble Alpes, Grenoble, 2018.

[MUL 16] MULLER A., VALETTE P.-J., Introduction à l'imagerie médicale : de la technique à la pratique, Lecture, INSA, February 2016.

[MUR 11] MURACCIOLI S., LAGIER V., Equipement de pointe : une gamma caméra CZT en médecine nucléaire, CHU Besançon, Besançon, 2011.

[MYE 96] MYERS W.G., "Georg Charles de Hevesy: The father of nuclear medicine". *J. Nucl. Med.* vol. 24, no. 4, 291–294, 1996.

[NAD 14] NADER M.B., Rôle du symporteur sodium iodure dans la carcinogenèse non thyroïdienne, PhD thesis, Université Paris Sud - Paris XI, Paris, 2014.

[NAL 14] NALDA E., CARRE A.L., PLISSONNEAU T., Guide à l'attention des prescripteurs du Service de Médecine Nucléaire, available at: www.chica-gap.fr, 2014.

[NDA 23] NDAR INFO, Cancer : 8 mille personnes tuées par an au Sénégal, available at: https://www.ndarinfo.com/Cancer-8-mille-personnes-tuees-par-an-au-Senegal_a35278.html, 2023.

[NEL 87] NELSON T.R., TUNG S.M., "Temperature dependence of proton relaxation times in vitro", *Magnetic Resonance Imaging*, vol. 5, no. 3, pp. 189–199, 1987.

[NER 09] NERISSON P., Modélisation du transfert des aérosols dans un local ventilé, PhD thesis, Université de Toulouse, Toulouse, 2009.

[NGU 23] N'GUESSAN K.J.-F., SAKHO I., KA K. et al., "Secondary cancer risk assessment after high-risk and intermediate-risk prostate cancer radiotherapy in Senegal", *World Journal of Advanced Research and Reviews*, vol. 17, pp. 240–248, 2023.

[NIC 10] NICOL S., Étude et construction d'un tomographe TEP/TDM pour petits animaux, combinant modules phoswich à scintillateurs et détecteur à pixels hybrides, PhD thesis, Université de la Méditerranée – Aix-Marseille II, Marseille, 2010.

[NIE 50] NIER A.O., "A redetermination of the relative abundances of the isotopes of carbon, nitrogen, oxygen, argon and potassium", *Physical Reviews*, vol. 77, pp. 789–793, 1950.

[NWA 19] NWATSOCK J.F., MBENG A.V., DONG À.-Z., "Indications et efficacité à 6 mois de l'irathérapie à activité ablative fixe dans l'hyperthyroïdie à Yaoundé", *Health Sciences and Disease*, vol. 20, pp. 31–36, 2019.

[ODO 18] ODOU M.-.F. Créatinine – créatininémie – créatininurie, available at: https://www.doctissimo.fr/html/sante/analyses/ana_equil_ions06.htm, 2018.

[OPE 07] OPECST, La médecine nucléaire, available at: https://www.senat.fr/opecst/note/2007_1_medecine_nucleaire_4p.pdf, 2007.

[PAL 07] PALLARDY G., PALLARDY M.-J., "Histoire abrégée du radiodiagnostic et de l'imagerie médicale", *Histoire des sciences médicales*, vol. XLI, no. 1, 2007.

[PAP 09] PAPASTEFANOU C., "Beryllium-7 aerosols in ambient air. Aerosol and air", *Quality Research*, vol. 9, no. 2, pp. 187–197, 2009.

[PAT 12] PATUREL G., L'effet Doppler Fizeau, available at: http://clea-astro.eu/lunap/DopplerFizeau, 2012.

[PAY 08] PAYOUK P., ALONSO M., ESQUERRÉ J.P. et al., "Les « nouveaux » radiopharmaceutiques", *Médecine Nucléaire*, vol. 32, no. 8, pp. 456–461, doi: 10.1016/j.mednuc.2008.06.007, 2008.

[PEL 18] PELLETIER A., La scintigraphie osseuse, available at: https://www.doctissimo.fr/scintigraphie-osseuse, 2018.

[PER 14] PERON-MAGNAN T., Le DaTSCAN, un marqueur pour le pronostic de la maladie de Parkinson ?, available at: http://thierryperonmagnan.unblog.fr/2014/09/02/le-datscan-un-marqueur-pour-le-pronostic-de-la-maladie-de-parkinson/, 2014.

[PET 14] PETITGUILLAUME A., Dosimétrie Monte Carlo personnalisée pour la planification et l'évaluation des traitements de radiothérapie interne : développement et application à la radiothérapie interne sélective (SIRT), PhD thesis, Université Paris Sud, Paris, 2014.

[POI 19] POIRIER-QUINOT M., "L'imagerie médicale, quand la physique rencontre la médecine", *Culture Science de l'Ingénieure*, École Normale Supérieure Paris-Saclay, Paris, 2019.

[POI 20] POIRIER S., Position du corps et déposition pulmonaire lors de la thérapie inhalée par nébulisation : revue de littérature, Master's thesis, Aix-Marseille Université, Marseille, available at: https://dumas.ccsd.cnrs.fr/document, 2020.

[POS 85] POSIN J.P., ARAKAWA M., CROOKS L.E. et al., "Hydrogen MR imaging of the head at 0.35 T and 0.7 T: Effects of magnetic field strength", *Radiology*, vol. 157, no. 3, pp. 679–683, 1985.

[POT 55] POTRATZ H.A., *A Radio-Chemical Procedure for Thorium and its Application to the Determination of Ionium in Coral Limestone*, Los Alamos Scientific Laboratory of the University of California, Los Alamos, 1955.

[POT 14] POTTIER V., Étude du myélome multiple au travers de l'imagerie médicale, PhD thesis, Université de Bordeaux, Bordeaux, 2014.

[PUR 46] PURCELL E.M., TORREY H.C., POUND R.V., "Resonance absorption by nuclear magnetic moments in a solid", *Physical Review*, vol. 69, p. 37, 1946.

[QAI 11] QAIM S.M., TARKANYI F., CAPOTE R., *Nuclear Data for the Production of Therapeutic Radionuclides*, IAEA, Vienna, available at: https://www-pub.iaea.org/MTCD/publications/PDF/trs473_web.pdf, 2011.

[RAD 21] RADIOLOGICAL SOCIETY OF NORTH AMERICA, Tomographie par émission de positons – Tomodensitométrie (TEP/CT), available at: https://www.radiologyinfo.org/en/info/pet, 2021.

[RAM 20] RAMSAY SANTÉ., Infarctus du myocarde, available at: https://ramsaygds.fr/infarctus-du-myocarde, 2020.

[RAT 19] RATHMANN S.M., AHMAD Z., SLIKBOER S. et al., "The radiopharmaceutical chemistry of technetium-99m", in LEWIS J.S., WINDHORST A.D., ZEGLIS B.M. (eds), *Radiopharmaceutical Chemistry*, Springer, Cham, pp. 311–333, doi: 10.1007/978-3-319-98947-1_18, 2019.

[REI 18] REINAUD F., Scintigraphie pulmonaire, available at: https://www.concilio.com/pneumologie-examens-scintigraphie-pulmonaire, 2018.

[REN 15] RENAUD P., Les essais atmosphériques d'armes nucléaires : des retombées radioactives à l'échelle planétaire, available at: https://www.irsn.fr/sites/default/files/retombees-tirs-armes-nucleaires/pdf, 2015.

[REN 16] RENARD J., Intérêt de l'imagerie par résonance magnétique (IRM) fonctionnelle (diffusion et spectroscopie) dans l'évaluation du tissu cérébral du Chien, Professional thesis, École Nationale Vétérinaire de Toulouse – ENVT, 2016.

[RIM 21] RIMBU A., VAN DE WIELE C., Scintigraphie parathyroïdienne, available at: http://www.nucleanord.fr/informations-scintigraphie-parathyroidienne-15.html, 2021.

[RIM 23] RIMBU A., VAN DE WIELE C., Scintigraphie de la perfusion cérébrale, available at: http://www.nucleanord.fr/scintigraphie-de-la-perfusion-cerebrale, 2023.

[ROG 82] ROGERS D.W.O., "More realistic Monte Carlo calculations of photon detector response functions", *Nuclear Instruments and Methods in Physics Research*, vol. 199, pp. 531–547, 1982.

[ROQ 13] ROQUEFEUIL M.S., Adaptation des paramètres temporels en imagerie par résonance magnétique en fonction des variations physiologiques du rythme cardiaque. Application à la cartographie T_2, PhD thesis, Université de Lorraine, Lorraine, 2013.

[ROS 17] ROSSANT-LUMBROSO J., ROSSANT L., Les accidents ischémiques transitoires (AIT) ou éclipses cérébrales, available at: https://www.doctissimo.fr/sante/avc/accidents-ischemiques-transitoires, 2017.

[ROS 21] ROSIER C., Une naine brune, nommée "L'accident", intrigue les astronomes, available at: https://www.rtbf.be/detail_une-naine-brune?id=1083692, 2021.

[ROS 22] ROSSANT-LUMBROSO J., ROSSANT L., La prostatite aiguë ou chronique : quels sont les symptômes ? Est-ce grave ?, available at: https://www.doctissimo.fr/html/sante/encyclopedie/sa_523_prostatite.htm, 2022.

[ROU 11] ROUVET J., Détermination de la pureté radiochimique de préparation radiopharmaceutiques par chromatographie liquide haute performance (HPLC) (mise en place du contrôle qualité du ^{99m}Tc-mertiatide (MAG3®) et étude de radiotraceurs innovants), PhD thesis, Université de Rouen, Rouen, 2011.

[ROU 14] ROUSSEL J., L'effet Doppler, available at: https://femto-physique.fr/presentation-doppler.pdf, 2014.

[ROU 17] ROUDIL J., Influence de la pathologie de Lewy sur le phénotype clinique de la maladie d'Alzheimer, PhD thesis, Université de Lorraine, Lorraine, 2017.

[RUI 13] RUIVO J.C., ZAMBONI C.B., MEDINA N.H. et al., "The half-life of ^{131g,m}Te", *AIP Conf. Proc.*, vol. 1529, p. 187, doi: 10.1063/1.4804114, 2013.

[SAB 19] SABA D., Les radionucléides et le plomb émis par une usine de fertilisants au Liban : étude de leurs mobilités dans les sols et les plantes pour une évaluation du risque sanitaire, PhD thesis, Université Paris-Saclay, Paris, 2019.

[SAK 96] SAKHO I., DIOP A., GAYE B. et al., "Chromatographie ionique avec détection conductimétrique. Application à la détermination quantitative des anions inorganiques dans les eaux de distribution urbaine du Sénégal", *2e journées annuelles de la société ouest africaine de chimie*, Dakar, 1996.

[SAK 97] SAKHO I., Détermination des anions inorganiques dans les eaux de consommation de trois régions du Sénégal (Dakar, Thiès et Diourbel) par la chromatographie ionique, Diploma of Advanced Studies in Physical Chemistry Applied to Energy, Université Cheikh Anta Diop de Dakar, Dakar, 1997.

[SAK 16] SAKHO I., *Guides Pratiques du Lycéen, Physique Terminales S*, Nouvelles Éditions Africaines du Sénégal, Dakar, 2016.

[SAK 17] SAKHO I., Interaction rayonnement-matière, Master's thesis, Université Assane Seck de Ziguinchor, Ziguinchor, 2017.

[SAK 18] SAKHO I., *Electrostatique, magnétostatique et induction électromagnétique, Résumés de cours & 120 exercices corrigés*, Éditions Ellipses, Paris, 2018.

[SAK 19] SAKHO I., *Introduction to Quantum Mechanics 1: Thermal Radiation and Experimental Facts Regarding the Quantization of Matter*, ISTE Ltd, London and John Wiley & Sons, New York, 2019.

[SAK 20a] SAKHO I., *Physique atomique : systèmes hydrogènoïdes et systèmes héliumoïdes, Cours & exercices corrigés*, Éditions Ellipses, Paris, 2020.

[SAK 20b] SAKHO I., *Introduction to Quantum Mechanics 2: Wave-Corpuscle, Quantization & Schrödinger's Equation*, ISTE Ltd, London and John Wiley & Sons, New York, 2020.

[SAK 22] SAKHO I., *Nuclear Physics 1: Nuclear Deexcitations, Spontaneous Nuclear Reactions*, ISTE Ltd, London and John Wiley & Sons, New York, 2022.

[SAK 23a] SAKHO I., FROMM M., MULLER D. et al., Quantification des dommages induits à l'AND par des électrons de basses énergies (0–20 eV), FARI 2019, Université de Thiès, Thiès, 2023.

[SAK 23b] SAKHO I., "Sakholian radius-to-mass ratio postulate applied to the calculation of the mass or the radius of a satellite in the solar system and in the milky way", *International Astronomy and Astrophysics Research Journal*, vol. 5, no. 1, pp. 189–200, available at: https://journaliaarj.com/index.php/IAARJ/article/view/95, 2023.

[SAL 20] SALAUN P.-Y., Instrumentation en médecine nucléaire, Médecine Nucléaire et Biophysique CHRU Brest – UBO, available at: https://cerf.radiologie.fr/, 2020.

[SAN 07] SANTOS F.O., Diffusion élastique résonante. Diffusion inélastique et réactions astrophysiques. GANIL (Grand Accélérateur National d'Ions Lourds), available at: https://www.osti.gov/etdeweb/, 2007.

[SAN 12] SANCHEZ-CABEZA J.A., RUIZ-FERNÁNDEZ A.C., "^{210}Pb sediment radiochronology : An integrated formulation and classification of dating models", *Geochimica et Cosmochimica Acta*, vol. 82, pp. 183–200, 2012.

[SAS 15] SASCO R., Développement d'un outil chronostratigraphique pour les archives climatiques : datations absolues (K/Ar, Ar/3Ar) et paléomagnétisme appliqués aux laves, PhD thesis, Université Paris Sud-Paris XI, Paris, 2015.

[SCA 11] SCACCHI G., Nucléosynthèse. Formation des éléments chimiques dans l'Univers. Vie et mort des étoiles ALS 8/12/2011, available at: http://als.univ-lorraine.fr/files/conferences/2011/nucleosynthese.pdf, 2011.

[SCH 07] SCHNEIDER J.A., ARVANITAKIS Z., BANG W. et al., "Mixed brain pathologies account for most dementia cases in community-dwelling older persons", *Neurology*, vol. 69, pp. 2197–204, doi: 10.1212/01.wnl.0000271090.28148.24, 2007.

[SCH 21] SCHER N., L'espérance de vie du cancer de la prostate de stade avancé, available at: https://radiotherapie-hartmann.fr/actualites/cancer-prostate/, 2021.

[SCH 22] SCHER N., Cancer de la prostate : les premiers symptômes d'alerte, available at: https://radiotherapie-hartmann.fr/cancer-de-la-prostate-les-premiers-symptomes-dalerte/, 2022.

[SÉG 15] SÉGALA G., L'angiogenèse du tissu tumoral, available at: https://www.futura-sciences.com/sante/, 2015.

[SÈN 19a] SÈNE M., NDEYE M., "Quantifying fossil fuel CO2 component over Dakar from 1960 to 2010 by radiocarbon observation in atmospheric CO2 and using Mauna Loa as background", *International Journal of Energy and Environmental Science*, vol. 4, no. 4, pp. 52–60, 2019.

[SÈN 19b] SÈNE M., NDEYE M., "La méthode du carbone 14 dans le monitoring de l'émission de dioxyde de carbone fossile dans l'atmosphère", *Bulletin IFAN. Dakar, Sér. A*, vol. 54, no. 1, pp. 117–136, 2019.

[SÈN 19c] SÈNE M., Le carbone 14 dans l'environnement. Contribution à l'étude de la pollution atmosphérique par le dioxyde de carbone dans les zones urbaines. Cas de la région de Dakar, PhD thesis, Université Cheikh Anta Diop de Dakar, Dakar, 2019.

[SEN 21] SENEWEB, Alzhimer : 1,5 million de sénégalais dans l'oubli, available at: https://www.seneweb.com/alzheimer-1-5-million-de-senegalais-dans-l-oubli_n_52565.html, 2021.

[SEP 08] SEPULVEDA A., SCHULLER P., WALLING D.E. et al., "Use of ^{7}Be to document soil erosion associated with a short period of extreme rainfall", *Journal of Environmental Radioactivity*, vol. 99, pp. 35–49, 2008.

[SHE 85] SHEN G.J., Datation des planchers stalagmitiques de sites acheuléens en Europe par les méthodes des déséquilibres des familles de l'uranium et contribution méthodologique, PhD thesis, Museum National d'Histoire Naturelle et Université Pierre et Marie Curie (Paris VI), Paris, 1985.

[SHP 74] SHPORER M.M., CIVAN M.M., "Effects of temperature and field strength on the NMR relaxation times of 23 Na in frog striated muscle", *Biochimica et Biophysica Acta*, vol. 354, no. 2, pp. 291–304, 1974.

[SOC 19a] SOCIÉTÉ FRANÇAISE DE MÉDECINE NUCLÉAIRE ET IMAGERIE MOLÉCULAIRE, La scintigraphie de la perfusion cérébrale, https://www.sfmn.org/206-la-scintigraphie-de-la-perfusion-cerebrale?showall=1, 2019.

[SOC 19b] SOCIÉTÉ FRANÇAISE DE MÉDECINE NUCLÉAIRE ET IMAGERIE MOLÉCULAIRE, La recherche du ganglion sentinelle, available at: https://www.sfmn.org/199-la-recherche-du-ganglion-sentinelle?start=4, 2019.

[SOC 21a] SOCIÉTÉ CANADIENNE DU CANCER, Chimiothérapie du cancer de la prostate, available at: https://cancer.ca/fr/cancer-information/cancer-types/prostate/treatment/chemotherapy, 2021.

[SOC 21b] SOCIÉTÉ FRANÇAISE D'ÉNERGIE NUCLÉAIRE, La médecine nucléaire, available at: https://www.sfen.org/energie-nucleaire/energie-service-progres/medecine-nucleaire, 2021.

[SOC 23] SOCIÉTÉ CANADIENNE DU CANCER, Prostatite, available at: https://cancer.ca/fr/cancer-types, 2023.

[SPA 12] SPASIC E., Dosimétrie in vivo intracavitaire basée sur la luminescence stimulée optiquement de l'Al_2O_3:C dédiée à la curiethérapie, PhD thesis, Université de Lorraine, Lorraine, 2012.

[STA 05] STANISZ G.J., ODROBINA E., PUN J. et al., "T_1, T_2 relaxation and magnetization transfer in tissue at 3T", *Magnetic Resonance in Medicine*, vol. 54, no. 3, pp. 507–512, 2005.

[STE 77] STEIGER R.H., JÄGER E., "Subcommission on geochronology: Convention on the use of decay constants in geo- and cosmochronology", *Earth and Planetary Science Letters*, vol. 5, pp. 320–324, 1977.

[STU 98] STUIVER M., REIMER P.J., BARD E. et al., "CALIB Rev 4.3 (Data set 2)", *Radiocarbon*, vol. 40, pp. 1041–1083, doi: 10.1017/S0033822200019123, 1998.

[TAI 08] TAILLET R., VILLAIN L., FEBVRE P., *Dictionnaire de physique*, Éditions De Boeck Université, Brussels, 2008.

[TAL 04] TALBOT J.N., Guide pour la rédaction de protocoles pour la tomographie par émission de positons (TEP) au [18F]-fludésoxyglucose ([18F]-FDG) en cancérologie, Société Française de Biophysique et de Médecine Nucléaire (SFBMN), Paris, available at: www.sfbmn.org, 2004.

[TAT 15] TATISCHEFF V., La nucléosynthèse stellaire, Lecture, 25e Festival d'Astronomie de Fleurance, available at: http://fermedesetoiles.com/documents/supports/la-nucleosynthese-stellaire.pdf, 2015.

[THI 06] THIRION M., PERCHERON S., MIRA J.-P., "ThyrotoxicoseThyrotoxicosis", *Réanimation*, vol. 15, no. 6, pp. 497–505, doi: 10.1016/j.reaurg.2006.10.002, 2006.

[THI 07] THIAM C.O., Dosimétrie en radiothérapie et curiethérapie par simulation Monte-Carlo GATE sur grille informatique, PhD thesis, Université Blaise Pascal – Clermont-Ferrand II, Clermont-Ferrand, 2007.

[THO 18] THOMAS M.J., Port du tablier plombe : étude dosimétrique, PhD thesis, Université Lyon 1, Lyon, 2018.

[THU 62] THURBER D.L., "Anomalous $^{234}U/^{238}U$ in nature", *Geophysical Research*, vol. 67, pp. 4518–4520, 1962.

[THU 63] THURBER D.L., Natural variation in the ratio $^{234}U/^{238}U$ and investigation of the potential of ^{234}U for pleistocene chronology, PhD thesis, University of Columbia, New York, 1963.

[TOU 01] TOUZERY C., Dosimétrie des explorations diagnostiques en médecine nucléaire, Report, Société Française de Physique Médicale, no. 19-2001, available at: http://www.sfpm.asso.fr, 2001.

[TOU 18] TOURARI S., Place de l'irathérapie dans le traitement de la maladie de Basedow : expérience du service d'endocrinologie de l'Hôpital Militaire Avicenne Marrakech, PhD thesis, Université Cadi Ayyad, Marrakech, 2018.

[TRE 18] TREUIL J.-P., L'idée du Big-Bang : histoire de la construction progressive du modèle standard de la cosmologie, available at: http://scienceinter.com/construction%20modele%20standard.pdf. 2018.

[TRE 22] TREMBLAY L., 10 – Spectres et diagramme HR, Worksheet solutions, Collège Mérici, Quebec, available at: https://physique.merici.ca/astro/chap10astsol.pdf, 2022.

[TYN 69] TYNDALL J., *Le son : cours expérimental fait à l'Institution Royale*, Gauthier-Villars, Paris, available at: https://www.abebooks.com, 1869.

[UNI 18] UNION NATIONALE DES ASSOCIATIONS FRANCE ALZHEIMER ET MALADIES APPARENTÉES, La maladie à corps de Lewy, available at: https://www.francealzheimer.org/2018/01/Brochure-Maladie-%C3%A0-corps-de-Lewy, 2018.

[VAC 18] VACHEZ Y., Troubles neuropsychiatriques de la maladie de Parkinson et stimulation haute fréquence du noyau subthalamique : approche préclinique chez le rat de l'hypothèse dopaminergique de l'apathie, PhD thesis, Université Grenoble Alpes, Grenoble, 2018.

[VAL 09] VALA C., Synthèse de groupements prosthétiques glucidiques : vers de nouveaux traceurs peptidiques pour l'imagerie par tomographie par emission de positons (TEP), PhD thesis, Université Henri Poincaré, Nancy-Université, 2009.

[VAN 10] VANGIONI É., "La nucléosynthèse primordiale : une fenêtre sur la physique des premières minutes de l'Univers", in AUDOUZE J. (ed.), *Le ciel à découvert*, Éditions CNRS, Paris, 2010.

[VER 11] VERGER A., Comparaison de tomoscintigraphies myocardiques de stress et de repos entre une caméra à semi-conducteurs et des caméras conventionnelles : influence du protocole utilisé, PhD thesis, Université Henry Poincaré Nancy I, Nancy, 2011.

[VET 23] VETOPSY, Modèle standard des particules. Comment les interactions sont portées par des particules ?, available at: http://www.vetopsy.fr/modele-standard-particules/interactions-fondamentales-particules.php, 2023.

[VID 23] VIDAL, Quels sont les symptômes du cancer de la prostate ?, available at: https://www.vidal.fr/maladies/cancers/cancer-prostate/symptomes.html, 2023.

[WAQ 17] WAQUIER D., La maladie de Basedow, qu'est-ce que c'est ?, available at: https://www.passeportsante.net/fr/Fiche.aspx?doc=maladie-de-basedow_pm, 2017.

[WAS 98] WASTIEL C., KOSINSKI M., "Contrôle de qualité des produits radiopharmaceutiques par chromatographie liquide HPLC", *Analusis magazine*, vol. 26, no. 2, 1998.

[WEM 08] WEMEAU J.L., DO CAO C., Cancers anaplasiques de la thyroïde, available at: www.orpha.net/data/patho/Pro/fr/CancersAnaplasiquesThyroide-FRfrPro86766v01.pdf, 2008.

[WEM 11] WEMEAU J.-L, SADOUL J.-L, D'HERBOMEZ M. et al., "Recommandations de la société française d'endocrinologie pour la prise en charge des nodules thyroïdiens", *La presse médicale*, vol. 40, pp. 793–826, 2011.

[WIT 03] WITKOWSKI N., *Une histoire sentimentale des sciences*, Éditions du Seuil, Paris, available at: https://sirius.nathan.fr/9782091723761/assets/chapitre-3-exercice-resolu/preview, 2003.

[WOR 23a] WORLD HEALTH ORGANIZATION, Dementia, available at: https://www.who.int/health-topics/dementia#tab=tab_1, 2023.

[WOR 23b] WORLD HEALTH ORGANIZATION, Diabetes, available at: https://www.who.int/health-topics/diabetes#tab=tab_1, 2023.

[WOR 23c] WORLD HEALTH ORGANIZATION. Parkinson disease, available at: https://www.who.int/news-room/fact-sheets/detail/parkinson-disease, 2023.

[YOU 20] YOUNES H., Détection d'aiguilles et de grains dans les images ultrasonores pour la curiethérapie dynamique. Imagerie médicale, PhD thesis, Université Grenoble Alpes, Grenoble, 2020.

[ZIM 06] ZIMMERMANN R., *La médecine nucléaire. La radioactivité au service du diagnostic et de la thérapie*, EDP Sciences, Les Ulis, available at: http://livre21.com/LIVREF/F5/F005146.pdf, 2006.

Index

E, F

G, H

I, K

L, M

N, O

P, Q

R, S

T, U

V, W

X, Z

Other titles from

ISTE

in

Waves

2024

RÉVEILLAC Jean-Michel
Synthesizers and Subtractive Synthesis 1: Theory and Overview
Synthesizers and Subtractive Synthesis 2: Application and Practice

2022

GONTRAND Christian
Electromagnetism: Links to Special Relativity

LACAZE Pierre Camille, LACROIX Jean-Christophe
Nanotechnology and Nanomaterials for Energy

RÉVEILLAC Jean-Michel
Recording and Voice Processing 1: History and Generalities
Recording and Voice Processing 2: Working in the Studio

ROYER Daniel, VALIER-BRASIER Tony
Elastic Waves in Solids 1: Propagation

SAKHO Ibrahima
Nuclear Physics 1: Nuclear Deexcitations, Spontaneous Nuclear Reactions

VALIER-BRASIER Tony, ROYER Daniel
Elastic Waves in Solids 2: Radiation, Scattering, Generation

2021

DAHOO Pierre-Richard, LAKHLIFI Azzedine
Infrared Spectroscopy of Symmetric and Spherical Top Molecules for Space Observation 1
(Infrared Spectroscopy Set – Volume 3)
Infrared Spectroscopy of Symmetric and Spherical Top Molecules for Space Observation 2
(Infrared Spectroscopy Set – Volume 4)

2020

DANIELE Vito G., LOMBARDI Guido
Scattering and Diffraction by Wedges 1: The Wiener-Hopf Solution - Advanced Applications
(Waves and Scattering Set – Volume 1)
Scattering and Diffraction by Wedges 2: The Wiener-Hopf Solution - Advanced Applications
(Waves and Scattering Set – Volume 2)

SAKHO Ibrahima
Introduction to Quantum Mechanics 2: Wave-Corpuscle, Quantization & Schrödinger's Equation

2019

BERTRAND Pierre, DEL SARTO Daniele, GHIZZO Alain
The Vlasov Equation 1: History and General Properties

DAHOO Pierre-Richard, LAKHLIFI Azzedine
Infrared Spectroscopy of Triatomics for Space Observation
(Infrared Spectroscopy Set – Volume 2)

RÉVEILLAC Jean-Michel
Electronic Music Machines: The New Musical Instruments

ROMERO-GARCIA Vicente, HLADKY-HENNION Anne-Christine
Fundamentals and Applications of Acoustic Metamaterials: From Seismic to Radio Frequency
(Metamaterials Applied to Waves Set – Volume 1)

SAKHO Ibrahima
Introduction to Quantum Mechanics 1: Thermal Radiation and Experimental Facts Regarding the Quantization of Matter

2018

SAKHO Ibrahima
Screening Constant by Unit Nuclear Charge Method: Description and Application to the Photoionization of Atomic Systems

2017

DAHOO Pierre-Richard, LAKHLIFI Azzedine
Infrared Spectroscopy of Diatomics for Space Observation
(Infrared Spectroscopy Set – Volume 1)

PARET Dominique, HUON Jean-Paul
Secure Connected Objects

PARET Dominque, SIBONY Serge
Musical Techniques: Frequencies and Harmony

RÉVEILLAC Jean-Michel
Analog and Digital Sound Processing

STAEBLER Patrick
Human Exposure to Electromagnetic Fields

2016

ANSELMET Fabien, MATTEI Pierre-Olivier
Acoustics, Aeroacoustics and Vibrations

BAUDRAND Henri, TITAOUINE Mohammed, RAVEU Nathalie
The Wave Concept in Electromagnetism and Circuits: Theory and Applications

PARET Dominique
Antennas Designs for NFC Devices

PARET Dominique
Design Constraints for NFC Devices

WIART Joe
Radio-Frequency Human Exposure Assessment

2015

PICART Pascal
New Techniques in Digital Holography

2014

APPRIOU Alain
Uncertainty Theories and Multisensor Data Fusion

JARRY Pierre, BENEAT Jacques N.
RF and Microwave Electromagnetism

LAHEURTE Jean-Marc
UHF RFID Technologies for Identification and Traceability

SAVAUX Vincent, LOUËT Yves
MMSE-based Algorithm for Joint Signal Detection, Channel and Noise Variance Estimation for OFDM Systems

THOMAS Jean-Hugh, YAAKOUBI Nourdin
New Sensors and Processing Chain

TING Michael
Molecular Imaging in Nano MRI

VALIÈRE Jean-Christophe
Acoustic Particle Velocity Measurements using Laser: Principles, Signal Processing and Applications

VANBÉSIEN Olivier, CENTENO Emmanuel
Dispersion Engineering for Integrated Nanophotonics

2013

BENMAMMAR Badr, AMRAOUI Asma
Radio Resource Allocation and Dynamic Spectrum Access

BOURLIER Christophe, PINEL Nicolas, KUBICKÉ Gildas
Method of Moments for 2D Scattering Problems: Basic Concepts and Applications

GOURE Jean-Pierre
Optics in Instruments: Applications in Biology and Medicine

LAZAROV Andon, KOSTADINOV Todor Pavlov
Bistatic SAR/GISAR/FISAR Theory Algorithms and Program Implementation

LHEURETTE Eric
Metamaterials and Wave Control

PINEL Nicolas, BOURLIER Christophe
Electromagnetic Wave Scattering from Random Rough Surfaces: Asymptotic Models

SHINOHARA Naoki
Wireless Power Transfer via Radiowaves

TERRE Michel, PISCHELLA Mylène, VIVIER Emmanuelle
Wireless Telecommunication Systems

2012

LALAUZE René
Chemical Sensors and Biosensors

LE MENN Marc
Instrumentation and Metrology in Oceanography

LI Jun-chang, PICART Pascal
Digital Holography

2011

BECHERRAWY Tamer
Mechanical and Electromagnetic Vibrations and Waves

BESNIER Philippe, DÉMOULIN Bernard
Electromagnetic Reverberation Chambers

GOURE Jean-Pierre
Optics in Instruments

GROUS Ammar
Applied Metrology for Manufacturing Engineering

LE CHEVALIER François, LESSELIER Dominique, STARAJ Robert
Non-standard Antennas

2010

BEGAUD Xavier
Ultra Wide Band Antennas

MARAGE Jean-Paul, MORI Yvon
Sonar and Underwater Acoustics

2009

BOUDRIOUA Azzedine
Photonic Waveguides

BRUNEAU Michel, POTEL Catherine
Materials and Acoustics Handbook

DE FORNEL Frédérique, FAVENNEC Pierre-Noël
Measurements using Optic and RF Waves

FRENCH COLLEGE OF METROLOGY
Transverse Disciplines in Metrology

2008

FILIPPI Paul J.T.
Vibrations and Acoustic Radiation of Thin Structures

LALAUZE René
Physical Chemistry of Solid-Gas Interfaces

2007

KUNDU Tribikram
Advanced Ultrasonic Methods for Material and Structure Inspection

PLACKO Dominique
Fundamentals of Instrumentation and Measurement

RIPKA Pavel, TIPEK Alois
Modern Sensors Handbook

2006

BALAGEAS Daniel *et al.*
Structural Health Monitoring

BOUCHET Olivier *et al.*
Free-Space Optics

BRUNEAU Michel, SCELO Thomas
Fundamentals of Acoustics

FRENCH COLLEGE OF METROLOGY
Metrology in Industry

GUILLAUME Philippe
Music and Acoustics

GUYADER Jean-Louis
Vibration in Continuous Media